HOLT
SOCIAL
STUDIES

Introduction to Geography

Christopher L. Salter

HOLT, RINEHART AND WINSTON

A Harcourt Education Company

Orlando • **Austin** • New York • San Diego • Toronto • London

Author

Dr. Christopher L. Salter

Dr. Christopher L. "Kit" Salter is Professor Emeritus of geography and former Chair of the Department of Geography at the University of Missouri. He did his undergraduate work at Oberlin College and received both his M.A. and Ph.D. degrees in geography from the University of California at Berkeley.

Dr. Salter is one of the country's leading figures in geography education. In the 1980s he helped found the national Geographic Alliance network to promote geography education in all 50 states. In the 1990s Dr. Salter was Co-Chair of the National Geography Standards Project, a group of distinguished geographers who created *Geography for Life* in 1994, the document outlining national standards in geography. In 1990 Dr. Salter received the National Geographic Society's first-ever Distinguished Geography Educator Award. In 1992 he received the George Miller Award for distinguished service in geography education from the National Council for Geographic Education.

Over the years, Dr. Salter has written or edited more than 150 articles and books on cultural geography, China, field work, and geography education. His primary interests lie in the study of the human and physical forces that create the cultural landscape, both nationally and globally.

ISBN 0-03-043604-4

3 4 5 6 7 8 9 032 13 12 11 10 09 08 07

Reviewers

Academic Reviewers

Elizabeth Chako, Ph.D.
Department of Geography
The George Washington
    University

Altha J. Cravey, Ph.D.
Department of Geography
University of North Carolina

Eugene Cruz-Uribe, Ph.D.
Department of History
Northern Arizona University

Toyin Falola, Ph.D.
Department of History
University of Texas

Sandy Freitag, Ph.D.
Director, Monterey Bay History
    and Cultures Project
Division of Social Sciences
University of California,
    Santa Cruz

Oliver Froehling, Ph.D.
Department of Geography
University of Kentucky

Reuel Hanks, Ph.D.
Department of Geography
Oklahoma State University

Phil Klein, Ph.D.
Department of Geography
University of Northern Colorado

B. Ikubolajeh Logan, Ph.D.
Department of Geography
Pennsylvania State University

Marc Van De Mieroop, Ph.D.
Department of History
Columbia University
New York, New York

Christopher Merrett, Ph.D.
Department of History
Western Illinois University

Jesse P. H. Poon, Ph.D.
Department of Geography
University at Buffalo–SUNY

Thomas R. Paradise, Ph.D.
Department of Geosciences
University of Arkansas

Robert Schoch, Ph.D.
CGS Division of Natural Science
Boston University

Derek Shanahan, Ph.D.
Department of Geography
Millersville University
Millersville, Pennsylvania

David Shoenbrun, Ph.D.
Department of History
Northwestern University
Evanston, Illinois

Sean Terry, Ph.D.
Department of Interdisciplinary
    Studies, Geography and
    Environmental Studies
Drury University
Springfield, Missouri

Educational Reviewers

Carla Freel
Hoover Middle School
Merced, California

Dennis Neel Durbin
Dyersburg High School
Dyersburg, Tennessee

Tina Nelson
Deer Park Middle School
Randallstown, Maryland

Don Polston
Lebanon Middle School
Lebanon, Indiana

Robert Valdez
Pioneer Middle School
Tustin, California

Teacher Review Panel

Heather Green
LaVergne Middle School
LaVergne, Tennessee

John Griffin
Wilbur Middle School
Wichita, Kansas

Rosemary Hall
Derby Middle School
Birmingham, Michigan

Rose King
Yeatman-Liddell School
St. Louis, Missouri

Mary Liebl
Wichita Public Schools USD 259
Wichita, Kansas

Jennifer Smith
Lake Wood Middle School
Overland Park, Kansas

Melinda Stephani
Wake County Schools
Raleigh, North Carolina

Introduction to Geography

Reading Social Studies H1

Social Studies and Academic Words H4

Geography and Map Skills H5

Making This Book Work for You H18

Scavenger Hunt H20

Introduction to Geography 1

CHAPTER 1 **A Geographer's World** 2

Geography's Impact Video Series
Impact of Studying Geography

Section 1 Studying Geography 4
Section 2 Geography Themes and Essential Elements 10
Social Studies Skills Analyzing Satellite Images 15
Section 3 The Branches of Geography 16
Chapter Review 21
Standardized Test Practice 23

CHAPTER 2 **Planet Earth** 24

Geography's Impact Video Series
Impact of Water on Earth

Section 1 Earth and the Sun's Energy 26
Section 2 Water on Earth 30
Section 3 The Land 35
Case Study The Ring of Fire 42
Social Studies Skills Using a Physical Map 44
Chapter Review 45
Standardized Test Practice 47

Plate B

Plate A

magma

CHAPTER 3 **Climate, Environment, and Resources** 48

🌐 **Geography's Impact Video Series**
Impact of Weather

Section 1 Weather and Climate 50

Section 2 World Climates 55

Section 3 Natural Environments 62

Geography and History Earth's Changing Environments 66

Section 4 Natural Resources 68

Literature The River 73

Social Studies Skills Analyzing a Bar Graph 74

Chapter Review 75

Standardized Test Practice 77

CHAPTER 4 **The World's People** 78

🌐 **Geography's Impact Video Series**
Impact of Culture

Section 1 Culture 80

Section 2 Population 86

Section 3 Government and Economy 91

Social Studies Skills Organizing Information 96

Section 4 Global Connections 97

Chapter Review 101

Standardized Test Practice 103

Reference

Reading Social Studies

Using Prior Knowledge 106

Using Word Parts 107

Understanding Cause and Effect 108

Understanding Main Ideas 109

Atlas

United States: Physical 110

United States: Political 112

World: Physical 114

World: Political 116

North America: Physical 118

North America: Political 119

South America: Physical 120

South America: Political 121

Europe: Physical 122

Europe: Political 123

Asia: Physical 124

Asia: Political 125

Africa: Physical 126

Africa: Political 127

The Pacific: Political 128

The North Pole 129

The South Pole 129

Facts about the World 130

Gazetteer 134

English and Spanish Glossary 136

Economics Handbook 142

Index 144

Features

Case Study

Take a detailed look at important topics in geography.

The Ring of Fire 42

Geography and History

Explore the connections between the world's places and the past.

Earth's Changing Environments 66

Close-up

See how people live and what places look like by taking a close-up view of geography.

Interactive The Five Themes
of Geography 11

Interactive The Water Cycle............... 32

A Forest Ecosystem......................... 63

A Global Economy.......................... 98

Focus on Culture

Learn about some of the world's fascinating cultures.

The Midnight Sun 29

The Tuareg of the Sahara................... 58

CONNECTING TO . . .

Explore the connections between geography and other subjects.

TECHNOLOGY
Computer Mapping.......................... 19

SCIENCE
Soil Factory 64

MATH
Calculating Population Density............. 88

Satellite View

See the world through satellite images and explore what these images reveal.

The World 1

True Color Satellite Image of Italy 15

Infrared Satellite Image of Italy............. 15

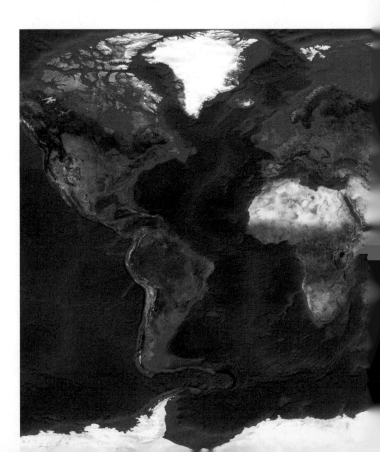

Social Studies Skills

Learn, practice, and apply the skills you need to study and analyze geography.

Analyzing Satellite Images 15

Using a Physical Map 44

Analyzing a Bar Graph 74

Organizing Information 96

Literature

Learn about the world's geography through literature.

The River 73

FOCUS ON READING

Learn and practice skills that will help you read your social studies lessons.

Using Prior Knowledge 106

Using Word Parts 107

Understanding Cause and Effect 108

Understanding Main Ideas 109

FOCUS ON WRITING AND VIEWING

Use writing and viewing skills to reflect on the world and its people.

Writing a Job Description 2

Writing a Haiku 24

Creating and Viewing
    a Weather Report 48

Creating a Poster 78

Explaining a Process 104

BIOGRAPHIES

Meet the people who have influenced the world and learn about their lives.

Eratosthenes 18

Alfred Wegener 37

Wangari Maathi 69

Primary Source

Learn about the world through important documents and personal accounts.

Geography for Life 14

Robert Heinlein, on climate, from
    Time Enough for Love 50

The Charter of the United Nations 100

Charts and Graphs

The *World Almanac and Book of Facts* is America's largest-selling reference book of all time, with more than 81 million copies sold since 1868.

FACTS ABOUT THE WORLD
Study the latest facts and figures about the world.

Eruptions in the Ring of Fire 42

World Energy Production Today 70

FACTS ABOUT COUNTRIES
Study the latest facts and figures about countries.

A Developed and a
Developing Country 94

Quick Facts and Infographics

Analyze visual information to learn about geography.

Geographic Dictionary..................... H14

What Is Geography? 5

Looking at the World 6

The Geographer's Tools 8

The Five Themes of Geography 11

Geography................................. 17

A Geographer's World 21

Solar Energy 27

The Seasons: Northern Hemisphere 28

Earth's Distribution of Water 30

The Water Cycle 32

Plate Movement........................... 38

Planet Earth 45

Global Wind Systems....................... 51

Rain Shadow Effect........................ 54

Highland Climates......................... 60

A Forest Ecosystem........................ 63

Soil Layers................................ 65

Climate, Environment, and Resources....... 75

A Global Economy......................... 98

The World's People........................ 101

Charts and Graphs

Use charts and graphs to analyze geographic information.

Percentage of Students on High School Soccer
Teams by Region 9

The Essential Elements and Geography
Standards 13

World Climate Regions..................... 56

Climate Graph for Nice, France.............. 59

Average Annual Precipitation
by Climate Region 74

Top Five Aluminum Producers, 2000........ 77

Irish Migration to the United States,
1845–1855 89

World Population Growth, 1500–2000...... 90

Economic Activity.......................... 93

Earth Facts................................ 130

World Population.......................... 132

Developed and Less Developed
Countries 132

World Religions............................ 133

World Languages.......................... 133

Geography Skills

Interactive Maps

Map Activity: Physical Map 46

World Climate Regions 56

Map Activity: Prevailing Winds 76

World Population Density 87

Map Activity: Population Density 102

Geography Skills With map zone geography skills, you can go online to find interactive versions of the key maps in this book. Explore these interactive maps to learn and practice important map skills and bring geography to life.

To use map zone interactive maps online:

1. Go to go.hrw.com.

2. Enter the KEYWORD shown on the interactive map.

3. Press return!

Maps

Northern Hemisphere H7

Southern Hemisphere H7

Western Hemisphere H7

Eastern Hemisphere H7

Mercator Projection H8

Conic Projection H9

Flat-plane Projection H9

The First Crusade, 1096 .... H10

Caribbean South America: Political H12

The Indian Subcontinent: Physical H13

West Africa: Climate H13

High School Soccer Participation 8

Sketch Map 22

The United States 23

Earth's Plates 36

Ring of Fire 42

India: Physical 44

Major Ocean Currents 52

Mediterranean Climate 59

Arab Culture Region 82

Cultural Diffusion of Baseball 84

Governments of the World 92

United States: Physical 110

United States: Political 112

World: Physical 114

World: Political 116

North America: Physical 118

North America: Political 119

South America: Physical 120

South America: Political 121

Europe: Physical 122

Europe: Political 123

Asia: Physical 124

Asia: Political 125

Africa: Physical 126

Africa: Political 127

The Pacific: Political 128

The North Pole 129

The South Pole 129

Become an Active Reader

by Dr. Kylene Beers

Did you ever think you would begin reading your social studies book by reading about *reading*? Actually, it makes better sense than you might think. You would probably make sure you knew some soccer skills and strategies before playing in a game. Similarly, you need to know something about reading skills and strategies before reading your social studies book. In other words, you need to make sure you know whatever you need to know in order to read this book successfully.

Tip #1

Read Everything on the Page!

You can't follow the directions on the cake-mix box if you don't know where the directions are! Cake-mix boxes always have directions on them telling you how many eggs to add or how long to bake the cake. But, if you can't find that information, it doesn't matter that it is there.

Likewise, this book is filled with information that will help you understand what you are reading. If you don't study that information, however, it might as well not be there. Let's take a look at some of the places where you'll find important information in this book.

The Chapter Opener
The chapter opener gives you a brief overview of what you will learn in the chapter. You can use this information to prepare to read the chapter.

The Section Openers
Before you begin to read each section, preview the information under What You Will Learn. There you'll find the main ideas of the section and key terms that are important in it. Knowing what you are looking for before you start reading can improve your understanding.

Boldfaced Words
Those words are important and are defined somewhere on the page where they appear— either right there in the sentence or over in the side margin.

Maps, Charts, and Artwork
These things are not there just to take up space or look good! Study them and read the information beside them. It will help you understand the information in the chapter.

Questions at the End of Sections
At the end of each section, you will find questions that will help you decide whether you need to go back and re-read any parts before moving on. If you can't answer a question, that is your cue to go back and re-read.

Questions at the End of the Chapter
Answer the questions at the end of each chapter, even if your teacher doesn't ask you to. These questions are there to help you figure out what you need to review.

Tip #2
Use the Reading Skills and Strategies in Your Textbook

Good readers use a number of skills and strategies to make sure they understand what they are reading. In this textbook you will find help with important reading skills and strategies such as "Using Prior Knowledge" and "Understanding Main Ideas."

We teach the reading skills and strategies in several ways. Use these activities and lessons and you will become a better reader.

- First, on the opening page of every chapter we identify and explain the reading skill or strategy you will focus on as you work through the chapter. In fact, these activities are called "Focus on Reading."

- Second, as you can see in the example at right, we tell you where to go for more help. The back of the book has a reading handbook with a full-page practice lesson to match the reading skill or strategy in every chapter.

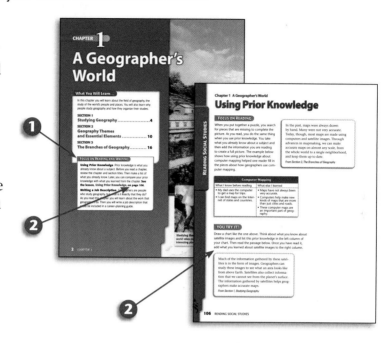

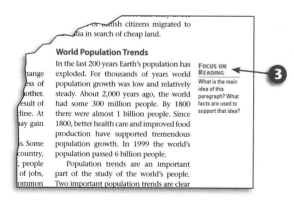

or British citizens migrated to ...lia in search of cheap land.

World Population Trends

In the last 200 years Earth's population has exploded. For thousands of years world population growth was low and relatively steady. About 2,000 years ago, the world had some 300 million people. By 1800 there were almost 1 billion people. Since 1800, better health care and improved food production have supported tremendous population growth. In 1999 the world's population passed 6 billion people.

Population trends are an important part of the study of the world's people. Two important population trends are clear

FOCUS ON READING

What is the main idea of this paragraph? What facts are used to support that idea?

- Third, we give you short practice activities and examples as you read the chapter. These activities and examples show up in the margin of your book. Again, look for the words, "Focus on Reading."

- Finally, we provide another practice activity in the Chapter Review at the end of every chapter. That activity gives you one more chance to make sure you know how to use the reading skill or strategy.

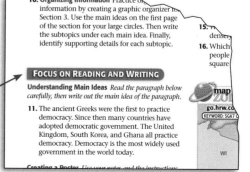

10. Organizing Information Practice o... information by creating a graphic organizer fo... Section 3. Use the main ideas on the first page of the section for your large circles. Then write the subtopics under each main idea. Finally, identify supporting details for each subtopic.

FOCUS ON READING AND WRITING

Understanding Main Ideas *Read the paragraph below carefully, then write out the main idea of the paragraph.*

11. The ancient Greeks were the first to practice democracy. Since then many countries have adopted democratic government. The United Kingdom, South Korea, and Ghana all practice democracy. Democracy is the most widely used government in the world today.

Tip #3

Pay Attention to Vocabulary

It is no fun to read something when you don't know what the words mean, but you can't learn new words if you only use or read the words you already know. In this book, we know we have probably used some words you don't know. But, we have followed a pattern as we have used more difficult words.

- First, at the beginning of each section you will find a list of key terms that you will need to know. Be on the lookout for those words as you read through the section. You will find that we have defined those words right there in the paragraph where they are used. Look for a word that is in boldface with its definition highlighted in yellow.

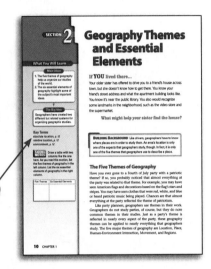

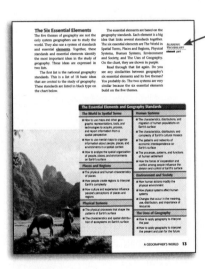

- Second, when we use a word that is important in all classes, not just social studies, we define it in the margin under the heading Academic Vocabulary. You will run into these academic words in other textbooks, so you should learn what they mean while reading this book.

Tip #4

Read Like a Skilled Reader

You won't be able to climb to the top of Mount Everest if you do not train! If you want to make it to the top of Mount Everest then you must start training to climb that huge mountain.

Training is also necessary to become a good reader. You will never get better at reading your social studies book—or any book for that matter—unless you spend some time thinking about how to be a better reader.

Skilled readers do the following:

1. They preview what they are supposed to read before they actually begin reading. When previewing, they look for vocabulary words, titles of sections, information in the margin, or maps or charts they should study.

2. They get ready to take some notes while reading by dividing their notebook paper into two parts. They title one side "Notes from the Chapter" and the other side "Questions or Comments I Have."

3. As they read, they complete their notes.

4. They read like **active readers**. The Active Reading list below shows you what that means.

5. Finally, they use clues in the text to help them figure out where the text is going. The best clues are called signal words. These are words that help you identify chronological order, causes and effects, or comparisons and contrasts.

Chronological Order Signal Words: *first, second, third, before, after, later, next, following that, earlier, subsequently, finally*

Cause and Effect Signal Words: *because of, due to, as a result of, the reason for, therefore, consequently, so, basis for*

Comparison/Contrast Signal Words: *likewise, also, as well as, similarly, on the other hand*

Active Reading

There are three ways to read a book: You can be a turn-the-pages-no-matter-what type of reader. These readers just keep on turning pages whether or not they understand what they are reading. Or, you can be a stop-watch-and-listen kind of reader. These readers know that if they wait long enough, someone will tell them what they need to know. Or, you can be an active reader. These readers know that it is up to them to figure out what the text means. Active readers do the following as they read:

Predict what will happen next based on what has already happened. When your predictions don't match what happens in the text, re-read the confusing parts.

Question what is happening as you read. Constantly ask yourself why things have happened, what things mean, and what caused certain events. Jot down notes about the questions you can't answer.

Summarize what you are reading frequently. Do not try to summarize the entire chapter! Read a bit and then summarize it. Then read on.

Connect what is happening in the section you're reading to what you have already read.

Clarify your understanding. Be sure that you understand what you are reading by stopping occasionally to ask yourself whether you are confused by anything. Sometimes you might need to re-read to clarify. Other times you might need to read further and collect more information before you can understand. Still other times you might need to ask the teacher to help you with what is confusing you.

Visualize what is happening in the text. In other words, try to see the events or places in your mind. It might help you to draw maps, make charts, or jot down notes about what you are reading as you try to visualize the action in the text.

Social Studies and Academic Words

As you read this textbook, you will be more successful if you know or learn the meanings of the words on this page. There are two types of words listed here. The first list contains social studies words. You will come across these words many times in your social studies classes. The second list contains academic words. These words are important in all of your classes, not just social studies. You will see these words in other textbooks, so you should learn what they mean while reading this book.

Social Studies Words

WORDS ABOUT TIME

AD	refers to dates after the birth of Jesus
BC	refers to dates before Jesus's birth
BCE	refers to dates before Jesus's birth, stands for "before the common era"
CE	refers to dates after Jesus's birth, stands for "common era"
century	a period of 100 years
decade	a period of 10 years
era	a period of time
millennium	a period of 1,000 years

WORDS ABOUT THE WORLD

climate	the weather conditions in a certain area over a long period of time
geography	the study of the world's people, places, and landscapes
physical features	features on Earth's surface, such as mountains and rivers
region	an area with one or more features that make it different from surrounding areas
resources	materials found on Earth that people need and value

WORDS ABOUT PEOPLE

anthropology	the study of people and cultures
archaeology	the study of the past based on what people left behind
citizen	a person who lives under the control of a government

civilization	the way of life of people in a particular place or time
culture	the knowledge, beliefs, customs, and values of a group of people
custom	a repeated practice or tradition
economics	the study of the production and use of goods and services
economy	any system in which people make and exchange goods and services
government	the body of officials and groups that run an area
history	the study of the past
politics	the process of running a government
religion	a system of beliefs in one or more gods or spirits
society	a group of people who share common traditions
trade	the exchange of goods or services

Academic Words

consequences	the effects of a particular event or events
distinct	clearly different and separate
element	the part of a particular event or events
factor	cause
innovation	a new idea or way of doing something
structure	the way something is set up or organized
traditional	customary, time-honored

Regions of
the World

Regions of the World

Geographers divide the world into regions for study. Each region has something about it that makes it unique and different from other regions. The map on the next page shows the major regions of the world. Explore this map to begin your study of geography.

How to Use the Map

The map on the next page is a special kind of map. It has transparent overlays that show different features and regions of the world. You can look at each overlay separately, or you can look at them together to see how they are connected. Just follow the steps below.

❶ **The Base Map** Start by lifting up all the transparent overlays and looking at only the base map. It shows the world's major oceans and seven continents, or large landmasses. What are the names of these continents? Where is each one located? Which oceans border each continent?

❷ **The First Overlay** Cover the base map with the first transparent overlay. It shows some of the world's major physical features, like rivers and mountains. First, study the rivers. Which rivers are shown? On which continents are they located? Next, look at the mountains. What mountain ranges can you see, and where are they?

❸ **The Second Overlay** Now cover the base map and first overlay with the second overlay. It shows the major regions of the world. The name of each region is listed at the bottom. Now, put it all together. What are the five regions shown? Which continents do they include, and where are they located? What are some major mountains and rivers found in each region? What oceans surround them? Finally, where are these major world regions located in relation to one another?

Regions of the World

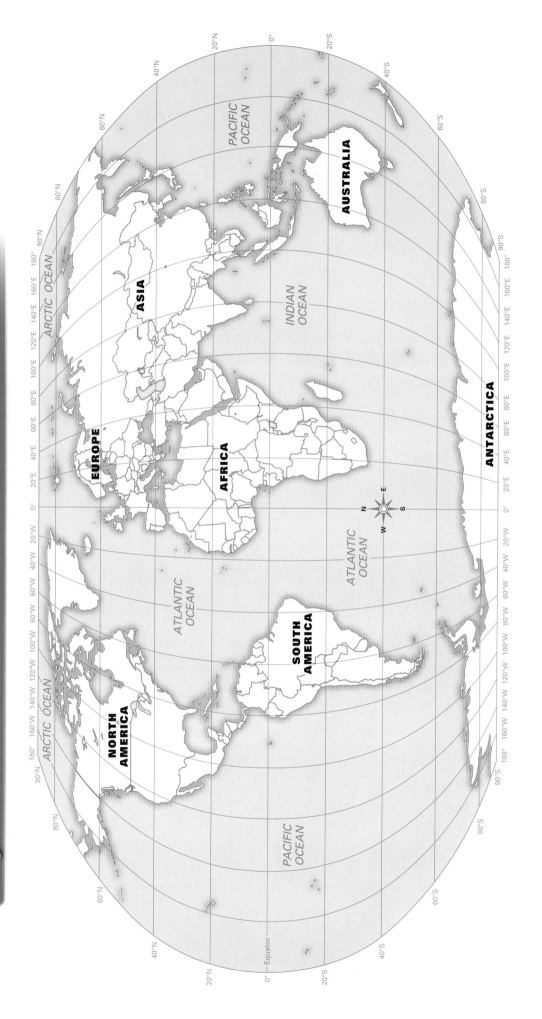

NORTH AMERICA

SOUTH AMERICA

EUROPE

ASIA

AFRICA

AUSTRALIA

ANTARCTICA

ARCTIC OCEAN

PACIFIC OCEAN

ATLANTIC OCEAN

INDIAN OCEAN

ATLANTIC OCEAN

PACIFIC OCEAN

Equator

0   1,000   2,000 Miles

0   1,000   2,000 Kilometers

Projection: Azimuthal Equal Area

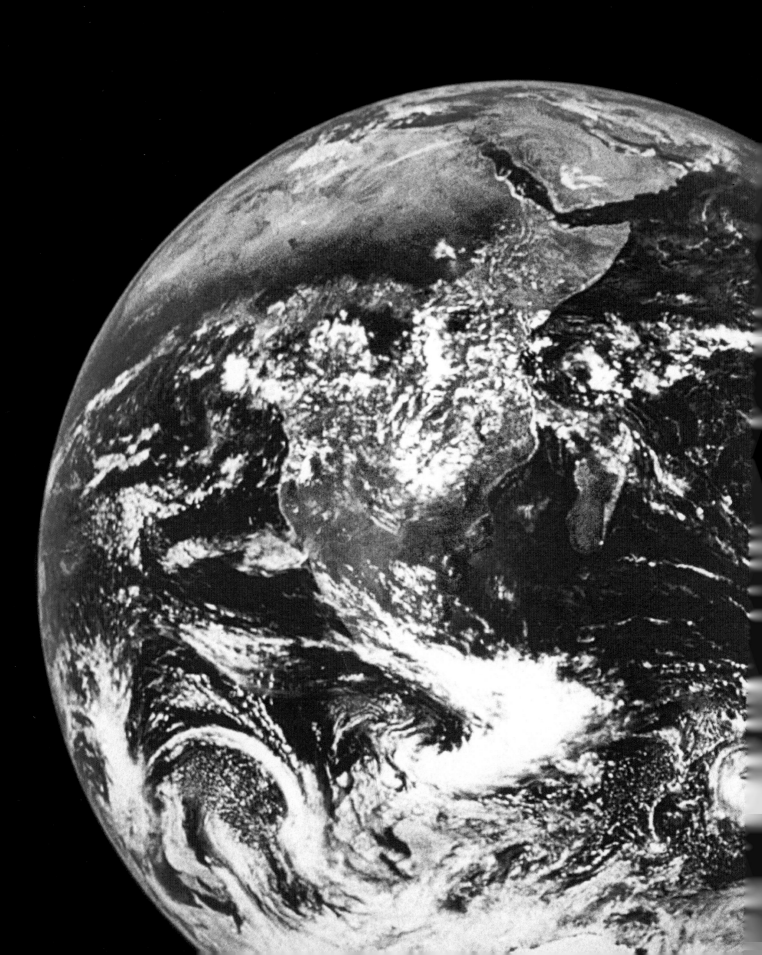

Geography and Map Skills Handbook

Contents

Mapping the Earth H6

Mapmaking H8

Map Essentials H10

Working with Maps H12

Geographic Dictionary H14

Themes and Essential Elements of Geography H16

Throughout this textbook, you will be studying the world's people, places, and landscapes. One of the main tools you will use is the map—the primary tool of geographers. To help you begin your studies, this Geography and Map Skills Handbook explains some of the basic features of maps. For example, it explains how maps are made, how to read them, and how they can show the round surface of Earth on a flat piece of paper. This handbook will also introduce you to some of the types of maps you will study later in this book. In addition, you will learn about the different kinds of features on Earth and about how geographers use themes and elements to study the world.

✷Interactive Maps

Geography Skills With map zone geography skills, you can go online to find interactive versions of the key maps in this book. Explore these interactive maps to learn and practice important map skills and bring geography to life.

To use map zone interactive maps online:

1. Go to go.hrw.com.
2. Enter the KEYWORD shown on the interactive map.
3. Press return!

Mapping the Earth
Using Latitude and Longitude

A **globe** is a scale model of the Earth. It is useful for showing the entire Earth or studying large areas of Earth's surface.

To study the world, geographers use a pattern of imaginary lines that circles the globe in east-west and north-south directions. It is called a **grid**. The intersection of these imaginary lines helps us find places on Earth.

The east-west lines in the grid are lines of **latitude**, which you can see on the diagram. Lines of latitude are called **parallels** because they are always parallel to each other. These imaginary lines measure distance north and south of the **equator**. The equator is an imaginary line that circles the globe halfway between the North and South Poles. Parallels measure distance from the equator in **degrees**. The symbol for degrees is °. Degrees are further divided into **minutes**. The symbol for minutes is ´. There are 60 minutes in a degree. Parallels north of the equator are labeled with an N. Those south of the equator are labeled with an S.

The north-south imaginary lines are lines of **longitude**. Lines of longitude are called **meridians**. These imaginary lines pass through the poles. They measure distance east and west of the **prime meridian**. The prime meridian is an imaginary line that runs through Greenwich, England. It represents 0° longitude.

Lines of latitude range from 0°, for locations on the equator, to 90°N or 90°S, for locations at the poles. Lines of longitude range from 0° on the prime meridian to 180° on a meridian in the mid-Pacific Ocean. Meridians west of the prime meridian to 180° are labeled with a W. Those east of the prime meridian to 180° are labeled with an E. Using latitude and longitude, geographers can identify the exact location of any place on Earth.

Lines of Latitude

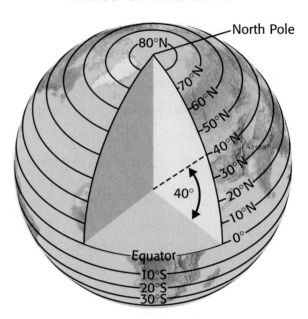

Lines of Longitude

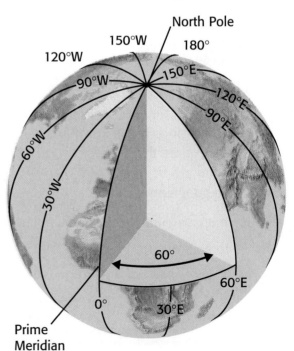

The equator divides the globe into two halves, called **hemispheres**. The half north of the equator is the Northern Hemisphere. The southern half is the Southern Hemisphere. The prime meridian and the 180° meridian divide the world into the Eastern Hemisphere and the Western Hemisphere. Look at the diagrams on this page. They show each of these four hemispheres.

Earth's land surface is divided into seven large landmasses, called **continents**. These continents are also shown on the diagrams on this page. Landmasses smaller than continents and completely surrounded by water are called **islands**.

Geographers organize Earth's water surface into major regions too. The largest is the world ocean. Geographers divide the world ocean into the Pacific Ocean, the Atlantic Ocean, the Indian Ocean, and the Arctic Ocean. Lakes and seas are smaller bodies of water.

Northern Hemisphere

Southern Hemisphere

Western Hemisphere

Eastern Hemisphere

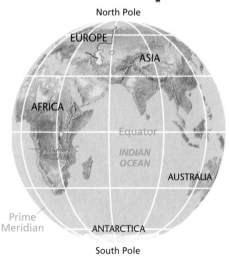

Mapmaking
Understanding Map Projections

A **map** is a flat diagram of all or part of Earth's surface. Mapmakers have created different ways of showing our round planet on flat maps. These different ways are called **map projections**. Because Earth is round, there is no way to show it accurately on a flat map. All flat maps are distorted in some way. Mapmakers must choose the type of map projection that is best for their purposes. Many map projections are one of three kinds: cylindrical, conic, or flat-plane.

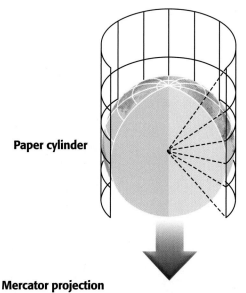

Paper cylinder

Mercator projection

Cylindrical Projections

Cylindrical projections are based on a cylinder wrapped around the globe. The cylinder touches the globe only at the equator. The meridians are pulled apart and are parallel to each other instead of meeting at the poles. This causes landmasses near the poles to appear larger than they really are. The map below is a Mercator projection, one type of cylindrical projection. The Mercator projection is useful for navigators because it shows true direction and shape. However, it distorts the size of land areas near the poles.

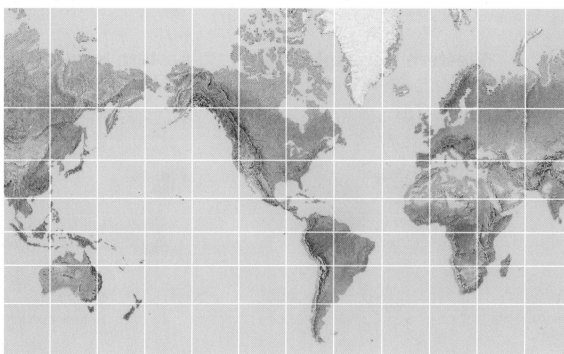

Conic Projections

Conic projections are based on a cone placed over the globe. A conic projection is most accurate along the lines of latitude where it touches the globe. It retains almost true shape and size. Conic projections are most useful for showing areas that have long east-west dimensions, such as the United States.

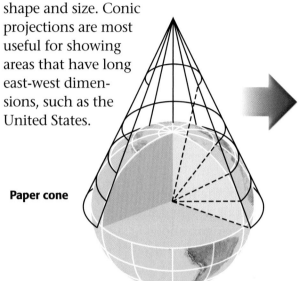

Paper cone

Conic projection

Flat-plane Projections

Flat-plane projections are based on a plane touching the globe at one point, such as at the North Pole or South Pole. A flat-plane projection is useful for showing true direction for airplane pilots and ship navigators. It also shows true area. However, it distorts the true shapes of landmasses.

Flat plane

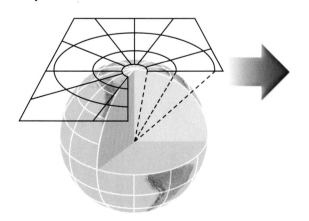

Flat-plane projection

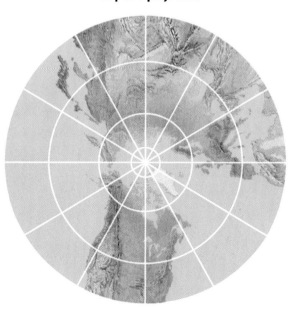

Map Essentials
How to Read a Map

Maps are like messages sent out in code. To help us translate the code, mapmakers provide certain features. These features help us understand the message they are presenting about a particular part of the world. Of these features, almost all maps have a title, a compass rose, a scale, and a legend. The map below has these four features, plus a fifth—a locator map.

❶ Title

A map's **title** shows what the subject of the map is. The map title is usually the first thing you should look at when studying a map, because it tells you what the map is trying to show.

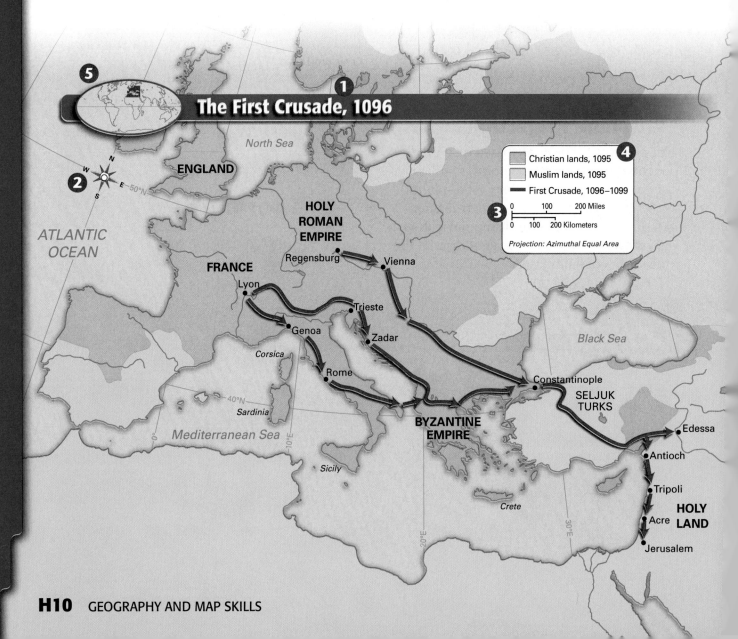

The First Crusade, 1096

Christian lands, 1095
Muslim lands, 1095
First Crusade, 1096–1099

0   100   200 Miles
0   100   200 Kilometers

Projection: Azimuthal Equal Area

North Sea
ENGLAND
ATLANTIC OCEAN
HOLY ROMAN EMPIRE
Regensburg
Vienna
FRANCE
Lyon
Trieste
Genoa
Zadar
Corsica
Rome
Black Sea
Sardinia
Constantinople
SELJUK TURKS
Mediterranean Sea
BYZANTINE EMPIRE
Edessa
Sicily
Antioch
Crete
Tripoli
HOLY LAND
Acre
Jerusalem

❷ Compass Rose

A directional indicator shows which way north, south, east, and west lie on the map. Some mapmakers use a "north arrow," which points toward the North Pole. Remember, "north" is not always at the top of a map. The way a map is drawn and the location of directions on that map depend on the perspective of the mapmaker. Most maps in this textbook indicate direction by using a compass rose. A **compass rose** has arrows that point to all four principal directions.

❸ Scale

Mapmakers use scales to represent the distances between points on a map. Scales may appear on maps in several different forms. The maps in this textbook provide a **bar scale**. Scales give distances in miles and kilometers.

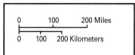

To find the distance between two points on the map, place a piece of paper so that the edge connects the two points. Mark the location of each point on the paper with a line or dot. Then, compare the distance between the two dots with the map's bar scale. The number on the top of the scale gives the distance in miles. The number on the bottom gives the distance in kilometers. Because the distances are given in large intervals, you may have to approximate the actual distance on the scale.

❹ Legend

The **legend**, or key, explains what the symbols on the map represent. Point symbols are used to specify the location of things, such as cities, that do not take up much space on the map. Some legends show colors that represent certain features like empires or other regions. Other maps might have legends with symbols or colors that represent features such as roads. Legends can also show economic resources, land use, population density, and climate.

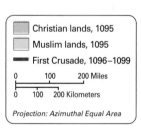

❺ Locator Map

A **locator map** shows where in the world the area on the map is located. The area shown on the main map is shown in red on the locator map. The locator map also shows surrounding areas so the map reader can see how the information on the map relates to neighboring lands.

Working with Maps
Using Different Kinds of Maps

As you study the world's regions and countries, you will use a variety of maps. Political maps and physical maps are two of the most common types of maps you will study. In addition, you will use special-purpose maps. These maps might show climate, population, resources, ancient empires, or other topics.

Political Maps

Political maps show the major political features of a region. These features include country borders, capital cities, and other places. Political maps use different colors to represent countries, and capital cities are often shown with a special star symbol.

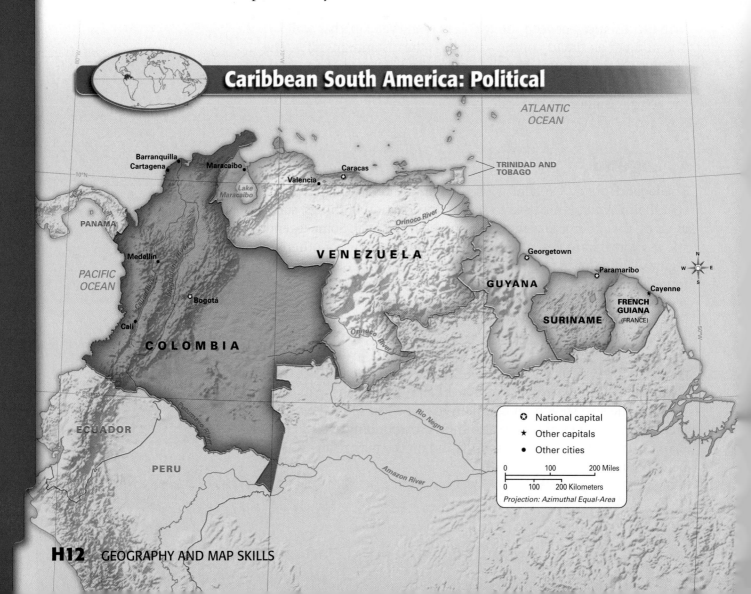

Caribbean South America: Political

ATLANTIC OCEAN

Barranquilla
Cartagena
Maracaibo
Caracas
Valencia
Lake Maracaibo
TRINIDAD AND TOBAGO
PANAMA
Orinoco River
VENEZUELA
Georgetown
Medellín
Paramaribo
PACIFIC OCEAN
GUYANA
Cayenne
Bogotá
FRENCH GUIANA (FRANCE)
SURINAME
Cali
Orinoco River
COLOMBIA
ECUADOR
Rio Negro
PERU
Amazon River

○ National capital
★ Other capitals
• Other cities

0   100   200 Miles
0 100   200 Kilometers
Projection: Azimuthal Equal-Area

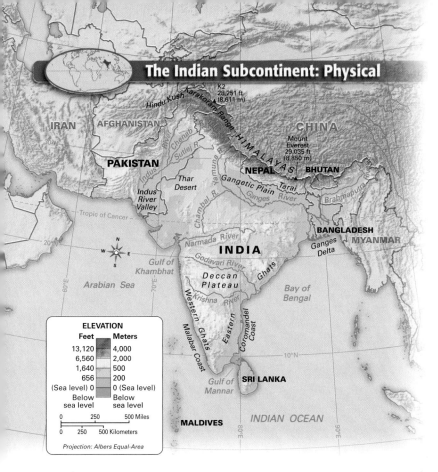

The Indian Subcontinent: Physical

K2 28,251 ft (8,611 m)

Mount Everest 29,035 ft (8,850 m)

IRAN
AFGHANISTAN
Hindu Kush
Karakoram Range
HIMALAYAS
CHINA
PAKISTAN
Chenab R.
Sutlej R.
Indus R.
Yamuna R.
NEPAL
BHUTAN
Thar Desert
Gangetic Plain
Tarai
Ganges
Brahmaputra
Indus River Valley
Tropic of Cancer
BANGLADESH
MYANMAR
Narmada River
INDIA
Ganges Delta
Gulf of Khambhat
Godavari River
Deccan Plateau
Ghats
Arabian Sea
Krishna River
Bay of Bengal
Western Ghats
Eastern Ghats
Coromandel Coast
Malabar Coast
10°N
Gulf of Mannar
SRI LANKA
MALDIVES
INDIAN OCEAN

ELEVATION

Feet	Meters
13,120	4,000
6,560	2,000
1,640	500
656	200
(Sea level) 0	0 (Sea level)
Below sea level	Below sea level

0   250   500 Miles
0   250   500 Kilometers

Projection: Albers Equal-Area

Physical Maps

Physical maps show the major physical features of a region. These features may include mountain ranges, rivers, oceans, islands, deserts, and plains. Often, these maps use different colors to represent different elevations of land. As a result, the map reader can easily see which areas are high elevations, like mountains, and which areas are lower.

Special-Purpose Maps

Special-purpose maps focus on one special topic, such as climate, resources, or population. These maps present information on the topic that is particularly important in the region. Depending on the type of special-purpose map, the information may be shown with different colors, arrows, dots, or other symbols.

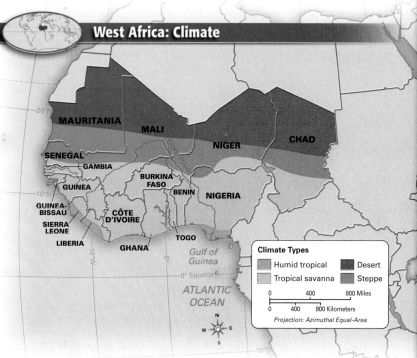

West Africa: Climate

MAURITANIA
MALI
NIGER
CHAD
SENEGAL
GAMBIA
GUINEA
BURKINA FASO
BENIN
NIGERIA
GUINEA-BISSAU
CÔTE D'IVOIRE
SIERRA LEONE
TOGO
LIBERIA
GHANA
Gulf of Guinea
0° Equator
ATLANTIC OCEAN

Climate Types

Humid tropical	Desert
Tropical savanna	Steppe

0   400   800 Miles
0   400   800 Kilometers

Projection: Azimuthal Equal-Area

Using Maps in Geography The different kinds of maps in this textbook will help you study and understand geography. By working with these maps, you will see what the physical geography of places is like, where people live, and how the world has changed over time.

Geographic Dictionary

OCEAN
a large body of water

CORAL REEF
an ocean ridge made up of
skeletal remains of tiny sea animals

GULF
a large part of
the ocean that
extends into land

PENINSULA
an area of land that sticks
out into a lake or ocean

BAY
part of a large
body of water
that is smaller
than a gulf

ISLAND
an area of land
surrounded entirely
by water

ISTHMUS
a narrow piece of land
connecting two larger
land areas

DELTA
an area where a
river deposits soil
into the ocean

STRAIT
a narrow body of
water connecting two
larger bodies of water

SINKHOLE
a circular depression
formed when the roof
of a cave collapses

WETLAND
an area of land
covered by
shallow water

RIVER
a natural flow of
water that runs
through the land

LAKE
an inland body
of water

FOREST
an area of densely
wooded land

COAST
an area of land
near the ocean

MOUNTAIN
an area of rugged
land that generally
rises higher than
2,000 feet

VALLEY
an area of low
land between
hills or mountains

GLACIER
a large area of
slow-moving ice

VOLCANO
an opening in Earth's crust
where lava, ash, and gases erupt

CANYON
a deep, narrow valley
with steep walls

HILL
a rounded, elevated
area of land smaller
than a mountain

PLAIN
a nearly
flat area

DUNE
a hill of sand
shaped by wind

OASIS
an area in the
desert with a
water source

DESERT
an extremely dry area with
little water and few plants

PLATEAU
a large, flat,
elevated
area of land

Themes and Essential Elements of Geography

by Dr. Christopher L. Salter

To study the world, geographers have identified 5 key themes, 6 essential elements, and 18 geography standards.

"How should we teach and learn about geography?" Professional geographers have worked hard over the years to answer this important question.

In 1984 a group of geographers identified the 5 Themes of Geography. These themes did a wonderful job of laying the groundwork for good classroom geography. Teachers used the 5 Themes in class, and geographers taught workshops on how to apply them in the world.

By the early 1990s, however, some geographers felt the 5 Themes were too broad. They created the 18 Geography Standards and the 6 Essential Elements. The 18 Geography Standards include more detailed information about what geography is, and the 6 Essential Elements are like a bridge between the 5 Themes and 18 Standards.

Look at the chart to the right. It shows how each of the 5 Themes connects to the Essential Elements and Standards. For example, the theme of Location is related to The World in Spatial Terms and the first three Standards. Study the chart carefully to see how the other themes, elements, and Standards are related.

The last Essential Element and the last two Standards cover The Uses of Geography. These key parts of geography were not covered by the 5 Themes. They will help you see how geography has influenced the past, present, and future.

5 Themes of Geography

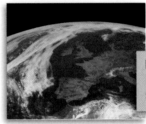

Location The theme of location describes where something is.

Place Place describes the features that make a site unique.

Regions Regions are areas that share common characteristics.

Movement This theme looks at how and why people and things move.

Human-Environment Interaction People interact with their environment in many ways.

6 Essential Elements

18 Geography Standards

I. The World in Spatial Terms

1. How to use maps and other tools
2. How to use mental maps to organize information
3. How to analyze the spatial organization of people, places, and environments

II. Places and Regions

4. The physical and human characteristics of places
5. How people create regions to interpret Earth
6. How culture and experience influence people's perceptions of places and regions

III. Physical Systems

7. The physical processes that shape Earth's surface
8. The distribution of ecosystems on Earth

IV. Human Systems

9. The characteristics, distribution, and migration of human populations
10. The complexity of Earth's cultural mosaics
11. The patterns and networks of economic interdependence on Earth
12. The patterns of human settlement
13. The forces of cooperation and conflict

V. Environment and Society

14. How human actions modify the physical environment
15. How physical systems affect human systems
16. The distribution and meaning of resources

VI. The Uses of Geography

17. How to apply geography to interpret the past
18. How to apply geography to interpret the present and plan for the future

Making This Book Work for You

Studying geography will be easy for you with this textbook. Take a few minutes now to become familiar with the easy-to-use structure and special features of your book. See how it will make geography come alive for you!

Your book begins with a satellite image of the world. You can explore this image to begin your study of geography.

Chapter

Each chapter includes an introduction, a Social Studies Skills activity, Chapter Review pages, and a Standardized Test Practice page.

Reading Social Studies Chapter reading lessons give you skills and practice to help you read the textbook. More help with each lesson can be found in the back of the book. Margin notes and questions in the chapter make sure you understand the reading skill.

Social Studies Skills The Social Studies Skills lessons give you an opportunity to learn, practice, and apply an important skill. Chapter Review questions then follow up on what you learned.

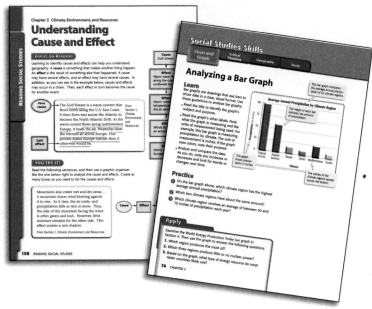

Section

The section opener pages include Main Ideas, an overarching Big Idea, and Key Terms and Places. In addition, each section includes these special features.

If YOU Lived There . . . Each section begins with a situation for you to respond to, placing you in a place that relates to the content you will be studying in the section.

Building Background Building Background connects what will be covered in each section with what you already know.

Short Sections of Content The information in each section is organized into small chunks of text that you can easily understand.

Taking Notes Suggested graphic organizers help you read and take notes on the important ideas in the section.

SECTION 1

Studying Geography

What You Will Learn...

Main Ideas
1. Geography is the study of the world, its people, and the landscapes they create.
2. Geographers look at the world in many different ways.
3. Maps and other tools help geographers study the planet.

The Big Idea
The study of geography and the use of geographic tools helps us view the world in new ways.

Key Terms
geography, p. 4
landscape, p. 4
social science, p. 5
region, p. 6
map, p. 8
globe, p. 8

TAKING NOTES Draw a large circle like the one below in your notebook. As you read this section, write a definition of geography at the top of the circle. Below that, list details about what geographers do.

Geography is ___

If YOU lived there...
You have just moved to Miami, Florida, from your old home in Pennsylvania. Everything seems very different—from the weather and the trees to the way people dress and talk. Even the streets and buildings look different. One day you get an e-mail from a friend at your old school. "What's it like living there?" he asks.

How will you describe your new home?

BUILDING BACKGROUND Often, when you are telling someone about a place they have never been, what you are describing is the place's geography. What the place looks like, what kind of weather it has, and how people live there are all parts of its geography.

What Is Geography?
Think about the place where you live. What does the land look like? Are there tall mountains nearby, or is the land so flat that you can see for miles? Is the ground covered with bright green grass and trees, or is the area part of a sandy desert?

Now think about the weather in your area. What is it like? Does it get really hot in the summer? Do you see snow every winter? How much does it rain? Do tornadoes ever strike?

Finally, think about the people who live in your town or city. Do they live mostly in apartments or houses? Do most people own cars, or do they get around town on buses or trains? What kinds of jobs do adults in your town have? Were most of the people you know born in your town, or did they move there?

The things that you have been thinking about are part of your area's geography. **Geography** is the study of the world, its people, and the landscapes they create. To a geographer, a place's **landscape** is all the human and physical features that make it unique. When they study the world's landscapes, geographers ask questions much like the ones you just asked yourself.

4 CHAPTER 1

Meteorology is the study of weather. This meteorologist is using computers to follow and predict the movement of a powerful storm.

Meteorology
Have you ever seen the ... television? If so, you ... of another branch ... branch called **meteorology**, the study of weather and what causes it.

Meteorologists study weather patterns in a particular area. Then they use the information to predict what the weather will be like in the coming days. Their work helps people plan what to wear and what to do on any given day. At the same time, their work can save lives by predicting the arrival of terrible storms. These predictions are among the most visible ways in which the work of geographers affects our lives every day.

READING CHECK Finding Main Ideas What are some major branches of geography?

SUMMARY AND PREVIEW In this section, you learned about two main branches of geography, physical and human. In the next chapter, you will learn more about the physical features that surround us and the processes that create them.

Hydrology

FOCUS ON READING
What do you already know about drinking water?

Another important branch of geography is hydrology, the study of water on Earth. Geographers in this field study the world's river systems and rainfall patterns. They study what causes floods and how people in cities can get safe drinking water. They also work to measure and protect the world's supply of water.

Section 3 Assessment

go.hrw.com
Online Quiz
KEYWORD: SG47 FP1

Reviewing Ideas, Terms, and Places
1. a. **Define** What is **physical geography**?
   b. **Explain** Why do we study physical geography?
2. a. **Identify** What are some things that people study as part of **human geography**?
   b. **Summarize** What are some ways in which the study of human geography can influence our lives?
   c. **Evaluate** Which do you think would be more interesting to study, physical geography or human geography? Why?
3. a. **Identify** What are two specialized fields of geography?
   b. **Analyze** How do cartographers contribute to the work of other geographers?

Critical Thinking
4. **Comparing and Contrasting** Draw a diagram like the one shown here. In the left circle, list three features of physical geography from your notes. In the right circle, list three features of human geography. Where the circles overlap, list one feature they share.

Physical   Human

FOCUS ON WRITING
5. **Choosing a Branch** Your job description should point out to people that there are many branches of geography. How will you note that?

20 CHAPTER 1

Reading Check Questions end each section of content so you can check to make sure you understand what you just studied.

Summary and Preview The Summary and Preview connects what you studied in the section to what you will study in the next section.

Section Assessment Finally, the section assessment boxes make sure that you understand the main ideas of the section. We also provide assessment practice online!

Scavenger Hunt

Are you ready to explore the world of geography? **_Holt Social Studies: Introduction to Geography_** is your ticket to this exciting world. Before you begin your journey, complete this scavenger hunt to get to know your book and discover what's inside.

On a separate sheet of paper, fill in the blanks to complete each sentence below. In each answer, one letter will be in a yellow box. When you have answered every question, copy these letters in order to reveal the answer to the question at the bottom of the page.

1 According to the Table of Contents, the title of Chapter 3 is Climate, Environment, and ⬜⬜⬜⬜⬜⬜⬜⬜. What else can you find in the Table of Contents?

7 The Close-up feature on page 98 is called A ⬜⬜⬜⬜⬜⬜ Economy. What other Close-up features can you find in the book?

2 Section 2 of Chapter 4 is called ⬜⬜⬜⬜⬜⬜⬜⬜⬜⬜⬜.

8 In the English and Spanish Glossary, the second word in the definition of *cartography* is ⬜⬜⬜⬜⬜⬜⬜.

3 Page 144 is the first page of the ⬜⬜⬜⬜⬜. How do you think you will use this section?

9 The Social Studies Skills lesson on page 107 is called Using ⬜⬜⬜⬜ ⬜⬜ ⬜⬜⬜⬜⬜.

4 The third key term listed on page 86 is ⬜⬜⬜⬜⬜⬜⬜⬜⬜.

10 The subject of the Writing Workshop on Page 104 is ⬜⬜⬜⬜⬜⬜⬜⬜⬜⬜⬜ a Process.

5 On page 35, the Main Ideas are followed by The ⬜⬜⬜ Idea. How do the Main Ideas connect to what is covered in this section?

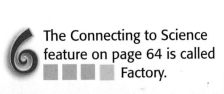

Fact!

About how many people live in the world today?

⬜⬜⬜ ⬜⬜⬜⬜⬜⬜⬜

6 The Connecting to Science feature on page 64 is called ⬜⬜⬜⬜ Factory.

Introduction to Geography

Deserts

Huge deserts, such as the Sahara in North Africa, are visible from space and appear yellow and brown.

Oceans

About 71 percent of Earth's surface is covered by vast amounts of salt water, which form the world's oceans.

Introduction to Geography

Explore the Satellite Image
Human-made machines that orbit Earth, called satellites, send back images of our planet like this one. What can you learn about Earth from studying this satellite image?

Frozen Lands

Earth's icy poles are frozen year-round and appear a brilliant white from space. These frozen lands contain much of Earth's freshwater.

A Geographer's World

What You Will Learn...

In this chapter you will learn about the field of geography, the study of the world's people and places. You will also learn why people study geography and how they organize their studies.

SECTION 1
Studying Geography **4**

SECTION 2
**Geography Themes
and Essential Elements** **10**

SECTION 3
The Branches of Geography **16**

FOCUS ON READING AND WRITING

Using Prior Knowledge Prior knowledge is what you already know about a subject. Before you read a chapter, review the chapter and section titles. Then make a list of what you already know. Later, you can compare your prior knowledge with what you learned from the chapter. **See the lesson, Using Prior Knowledge, on page 106.**

Writing a Job Description Geographers are people who study geography, but what is it exactly that they do? As you read this chapter you will learn about the work that geographers do. Then you will write a job description that could be included in a career-planning guide.

Studying the World Exploring the world takes people to exciting and interesting places.

ANALYSIS
SKILL **ANALYZING VISUALS**

This village is in the country of Nepal. It rests high in the Himalayas, the highest mountains in the world.

What is the land around the village like? How can you tell that people live in this area?

HOLT

Geography's Impact
video series
Watch the video to understand the impact of studying geography.

Human Geography Geography is also the study of people. It asks where people live, what they eat, what they wear, and even what kinds of animals they keep.

Physical Geography Geography is the study of the world's land features, such as this windswept rock formation in Arizona.

3

Studying Geography

What You Will Learn...

Main Ideas

1. Geography is the study of the world, its people, and the landscapes they create.
2. Geographers look at the world in many different ways.
3. Maps and other tools help geographers study the planet.

The Big Idea

The study of geography and the use of geographic tools helps us view the world in new ways.

Key Terms

geography, *p. 4*
landscape, *p. 4*
social science, *p. 5*
region, *p. 6*
map, *p. 8*
globe, *p. 8*

TAKING NOTES Draw a large circle like the one below in your notebook. As you read this section, write a definition of geography at the top of the circle. Below that, list details about what geographers do.

Geography is ___
_____

If YOU lived there...

You have just moved to Miami, Florida, from your old home in Pennsylvania. Everything seems very different—from the weather and the trees to the way people dress and talk. Even the streets and buildings look different. One day you get an e-mail from a friend at your old school. "What's it like living there?" he asks.

How will you describe your new home?

BUILDING BACKGROUND Often, when you are telling someone about a place they have never been, what you are describing is the place's geography. What the place looks like, what kind of weather it has, and how people live there are all parts of its geography.

What Is Geography?

Think about the place where you live. What does the land look like? Are there tall mountains nearby, or is the land so flat that you can see for miles? Is the ground covered with bright green grass and trees, or is the area part of a sandy desert?

Now think about the weather in your area. What is it like? Does it get really hot in the summer? Do you see snow every winter? How much does it rain? Do tornadoes ever strike?

Finally, think about the people who live in your town or city. Do they live mostly in apartments or houses? Do most people own cars, or do they get around town on buses or trains? What kinds of jobs do adults in your town have? Were most of the people you know born in your town, or did they move there?

The things that you have been thinking about are part of your area's geography. **Geography** is the study of the world, its people, and the landscapes they create. To a geographer, a place's **landscape** is all the human and physical features that make it unique. When they study the world's landscapes, geographers ask questions much like the ones you just asked yourself.

Geography as a Science

Many of the questions that geographers ask deal with how the world works. They want to know what causes mountains to form and what creates tornadoes. To answer questions like these, geographers have to think and act like scientists.

As scientists, geographers look at data, or information, that they gather about places. Gathering data can sometimes lead geographers to fascinating places. They might have to crawl deep into caves or climb tall mountains to make observations and take measurements. At other times, geographers study sets of images collected by satellites orbiting high above Earth.

However geographers gather their data, they have to study it carefully. Like other scientists, geographers must examine their findings in great detail before they can learn what all the information means.

Geography as a Social Science

Not everything that geographers study can be measured in numbers, however. Some geographers study people and their lives. For example, they may ask why countries change their governments or why people in a place speak a certain language. This kind of information cannot be measured.

Because it deals with people and how they live, geography is sometimes called a social science. A **social science** is a field that studies people and the relationships among them.

The geographers who study people do not dig in caves or climb mountains. Instead, they visit places and talk to the people who live there. They want to learn about people's lives and communities.

READING CHECK **Analyzing** In what ways is geography both a science and a social science?

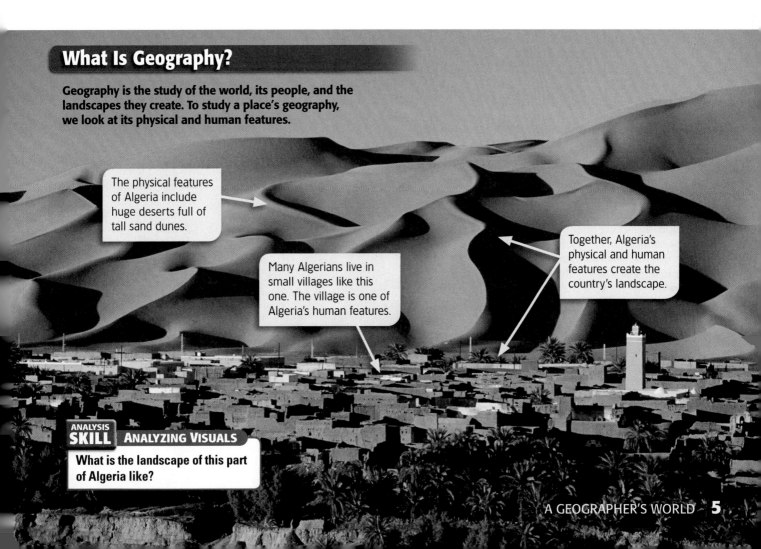

What Is Geography?

Geography is the study of the world, its people, and the landscapes they create. To study a place's geography, we look at its physical and human features.

The physical features of Algeria include huge deserts full of tall sand dunes.

Many Algerians live in small villages like this one. The village is one of Algeria's human features.

Together, Algeria's physical and human features create the country's landscape.

ANALYSIS SKILL **ANALYZING VISUALS**

What is the landscape of this part of Algeria like?

Looking at the World

Whether they study volcanoes and storms or people and cities, geographers have to look carefully at the world around them. To fully understand how the world works, geographers often look at places at three different levels.

Local Level

Some geographers study issues at a local level. They ask the same types of questions we asked at the beginning of this chapter: How do people in a town or community live? What is the local government like? How do the people who live there get around? What do they eat?

By asking these questions, geographers can figure out why people live and work the way they do. They can also help people improve their lives. For example, they can help town leaders figure out the best place to build new schools, shopping centers, or sports complexes. They can also help the people who live in the city or town plan for future changes.

Regional Level

Sometimes, though, geographers want to study a bigger chunk of the world. To do this, they divide the world into regions. A **region** is a part of the world that has one or more common features that distinguish it from surrounding areas.

Some regions are defined by physical characteristics such as mountain ranges, climates, or plants native to the area. As a result, these types of regions are often easy to identify. The Rocky Mountains of the western United States, for example, make up a physical region. Another example of this kind of region is the Sahara, a huge desert in northern Africa.

Other regions may not be so easy to define, however. These regions are based on the human characteristics of a place, such as language, religion, or history. A place in which most people share these kinds of characteristics can also be seen as a region. For example, most people in Scandinavia, a region in northern Europe, speak similar languages and practice the same religion.

Looking at the World

Geographers look at the world at many levels. At each level, they ask different questions and discover different types of information. By putting information gathered at different levels together, geographers can better understand a place and its role in the world.

ANALYZING VISUALS Based on these photos, what are some questions a geographer might ask about London?

Local Level This busy neighborhood in London, England, is a local area. A geographer here might study local foods, housing, or clothing.

Regions come in all shapes and sizes. Some are small, like the neighborhood called Chinatown in San Francisco. Other regions are huge, like the Americas. This huge region includes two continents, North America and South America. The size of the area does not matter, as long as the area shares some characteristics. These shared characteristics define the region.

Geographers divide the world into regions for many reasons. The world is a huge place and home to billions of people. Studying so large an area can be extremely difficult. Dividing the world into regions makes it easier to study. A small area is much easier to examine than a large area.

Other geographers study regions to see how people interact with one another. For example, they may study a city such as London, England, to learn how the city's people govern themselves. Then they can compare what they learn about one region to what they learn about another region. In this way, they can learn more about life and landscapes in both places.

Global Level

Sometimes geographers do not want to study the world just at a regional level. Instead they want to learn how people interact globally, or around the world. To do so, geographers ask how events and ideas from one region of the world affect people in other regions. In other words, they study the world on a global level.

Geographers who study the world on a global level try to find relationships among people who live far apart. They may, for example, examine the products that a country exports to see how those products are used in other countries.

In recent decades, worldwide trade and communication have increased. As a result, we need to understand how our actions affect people around the world. Through their studies, geographers provide us with information that helps us figure out how to live in a rapidly changing world.

READING CHECK **Finding Main Ideas** At what levels do geographers study the world?

Regional Level As a major city, London is also a region. At this level, a geographer might study the city's population or transportation systems.

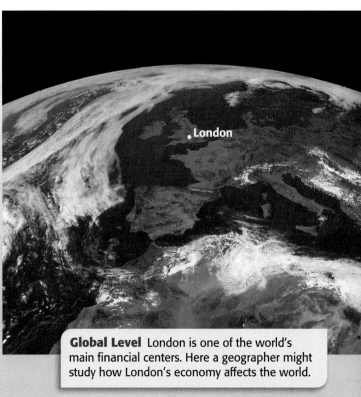

Global Level London is one of the world's main financial centers. Here a geographer might study how London's economy affects the world.

The Geographer's Tools

Geographers use many tools to study the world. Each tool provides part of the information a geographer needs to learn what a place is like.

ANALYZING VISUALS What information could you learn from each of these tools?

A geographer can use a globe to see where a place, such as the United States, is located.

Participation in High School Soccer

- More than 9%
- 5–9%
- 3–5%
- Fewer than 3%
- Data not available

Maps usually give geographers more information about a place than globes do. This map, for example, shows rates of soccer participation in the United States.

The Geographer's Tools

Have you ever seen a carpenter building or repairing a house? If so, you know that builders need many tools to do their jobs correctly. In the same way, geographers need many tools to study the world.

Maps and Globes

FOCUS ON READING

What do you already know about maps and globes?

The tools that geographers use most often in their work are maps and globes. A **map** is a flat drawing that shows all or part of Earth's surface. A **globe** is a spherical, or ball-shaped, model of the entire planet.

Both maps and globes show what the world looks like. They can show where mountains, deserts, and oceans are. They can also identify and describe the world's countries and major cities.

There are, however, major differences between maps and globes. Because a globe is spherical like Earth, it can show the world as it really is.

A map, though, is flat. It is not possible to show a spherical area perfectly on a flat surface. To understand what this means, think about an orange. If you took the peel off of an orange, could you make it lie completely flat? No, you could not, unless you stretched or tore the peel first.

The same principle is true with maps. To draw Earth on a flat surface, people have to distort, or alter, some details. For example, places on a map might look to be farther apart than they really are, or their shapes or sizes might be changed slightly.

Still, maps have many advantages over globes. Flat maps are easier to work with than globes. Also, it is easier to show small areas like cities on maps than on globes.

In addition, maps usually show more information than globes. Because globes are more expensive to make, they do not usually show anything more than where places are and what features they have.

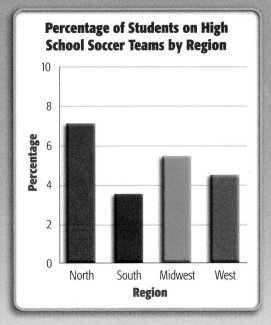

Percentage of Students on High School Soccer Teams by Region

Charts and graphs are also tools geographers can use to study information. They are often used when geographers want to compare numbers, such as the number of students who play soccer in each region of the country.

Maps, on the other hand, can show all sorts of information. Besides showing land use and cities, maps can include a great deal of information about a place. A map might show what languages people speak or where their ancestors came from. Maps like the one on the opposite page can even show how many students in an area play soccer.

Satellite Images

Maps and globes are not the only tools that geographers use in their work. As you have already read, many geographers study information gathered by satellites.

Much of the information gathered by these satellites is in the form of images. Geographers can study these images to see what an area looks like from above Earth. Satellites also collect information that we cannot see from the planet's surface. The information gathered by satellites helps geographers make accurate maps.

Other Tools

Geographers also use many other tools. For example, they use computer programs to create, update, and compare maps. They also use measuring devices to record data. In some cases, the best tools a geographer can use are a notebook and tape recorder to take notes while talking to people. Armed with the proper tools, geographers learn about the world's people and places.

READING CHECK **Summarizing** What are some of the geographer's basic tools?

SUMMARY AND PREVIEW Geography is the study of the world, its people, and its landscapes. In the next section, you will learn about two systems geographers use to organize their studies.

go.hrw.com
Online Quiz
KEYWORD: SGA7 HP1

Section 1 Assessment

Reviewing Ideas, Terms, and Places

1. **a. Define** What is **geography**?
   b. Explain Why is geography considered a science?
2. **a. Identify** What is a **region**? Give two examples.
   b. Elaborate What global issues do geographers study?
3. **a. Describe** How do geographers use satellite images?
   b. Compare and Contrast How are maps and globes similar? How are they different?

Critical Thinking

4. **Summarizing** Draw three ovals like the ones shown here. Use your notes to fill the ovals with information about geography, geographers, and their tools.

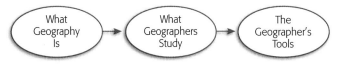

What Geography Is → What Geographers Study → The Geographer's Tools

FOCUS ON WRITING

5. **Describing the Subject** Based on what you have learned, what might attract people to work in geography? In your notebook, list some details about geography that might make people interested in working in the subject.

A GEOGRAPHER'S WORLD **9**

Geography Themes and Essential Elements

What You Will Learn...

Main Ideas

1. The five themes of geography help us organize our studies of the world.
2. The six essential elements of geography highlight some of the subject's most important ideas.

The Big Idea

Geographers have created two different but related systems for organizing geographic studies.

Key Terms

absolute location, *p. 12*
relative location, *p. 12*
environment, *p. 12*

TAKING NOTES Draw a table with two columns like the one here. As you read this section, list the five themes of geography in the left column. List the six essential elements of geography in the right column.

Five Themes	Six Essential Elements

If YOU lived there...

Your older sister has offered to drive you to a friend's house across town, but she doesn't know how to get there. You know your friend's street address and what the apartment building looks like. You know it's near the public library. You also would recognize some landmarks in the neighborhood, such as the video store and the supermarket.

What might help your sister find the house?

BUILDING BACKGROUND Like drivers, geographers have to know where places are in order to study them. An area's location is only one of the aspects that geographers study, though. In fact, it is only one of the five themes that geographers use to describe a place.

The Five Themes of Geography

Have you ever gone to a Fourth of July party with a patriotic theme? If so, you probably noticed that almost everything at the party was related to that theme. For example, you may have seen American flags and decorations based on the flag's stars and stripes. You may have seen clothes that were red, white, and blue or heard patriotic music being played. Chances are that almost everything at the party reflected the theme of patriotism.

Like party planners, geographers use themes in their work. Geographers do not study parties, of course, but they do note common themes in their studies. Just as a party's theme is reflected in nearly every aspect of the party, these geography themes can be applied to nearly everything that geographers study. The five major themes of geography are Location, Place, Human-Environment Interaction, Movement, and Regions.

Interactive Close-up

The Five Themes of Geography

Geographers use five major themes, or ideas, to organize and guide their studies.

go.hrw.com KEYWORD: SGA7 CH1

Location The theme of location describes where something is. The mountain shown above, Mount Rainier, is in west-central Washington.

Place Place describes the features that make a site unique. For example, Washington, D.C., is our nation's capital and has many great monuments.

UNITED STATES

Regions Regions are areas that share common characteristics. The Mojave Desert, shown here, is defined by its distinctive climate and plant life.

Movement This theme looks at how and why people and things move. Airports like this one in Dallas, Texas, help people move around the world.

Human-Environment Interaction People interact with their environments in many ways. Some, like this man in Florida, use the land to grow crops.

ANALYSIS SKILL ANALYZING VISUALS

Which of the five themes deals with the relationships between people and their surroundings?

A GEOGRAPHER'S WORLD **11**

Location

Every point on Earth has a location, a description of where it is. This location can be expressed in many ways. Sometimes a site's location is expressed in specific, or absolute, terms, such as an address. For example, the White House is located at 1600 Pennsylvania Avenue in the city of Washington, D.C. A specific description like this one is called an **absolute location**. Other times, the site's location is expressed in general terms. For example, Canada is north of the United States. This general description of where a place lies is called its **relative location**.

Place

Another theme, Place, is closely related to Location. However, Place does not refer simply to where an area is. It refers to the area's landscape, the features that define the area and make it different from other places. Such features could include land, climate, and people. Together, they give a place its own character.

Human-Environment Interaction

FOCUS ON READING
What do you know about environments?

In addition to looking at the features of places, geographers examine how those features interact. In particular, they want to understand how people interact with their environment—how people and their physical environment affect each other. An area's **environment** includes its land, water, climate, plants, and animals.

People interact with their environment every day in all sorts of ways. They clear forests to plant crops, level fields to build cities, and dam rivers to prevent floods. At the same time, physical environments affect how people live. People in cold areas, for example, build houses with thick walls and wear heavy clothing to keep warm. People who live near oceans look for ways to protect themselves from storms.

Movement

People are constantly moving. They move within cities, between cities, and between countries. Geographers want to know why and how people move. For example, they ask if people are moving to find work or to live in a more pleasant area. Geographers also study the roads and routes that make movement so common.

Regions

You have already learned how geographers divide the world into many regions to help the study of geography. Creating regions also makes it easier to compare places. Comparisons help geographers learn why each place has developed the way it has.

READING CHECK Finding Main Ideas What are the five themes of geography?

The six essential elements are used by geographers to organize their studies and are closely related to the geography standards. Each element includes several of the standards, as listed in the chart.

ANALYZING VISUALS How many of the six essential elements can you see illustrated in this photo?

The Six Essential Elements

The five themes of geography are not the only system geographers use to study the world. They also use a system of standards and essential <u>elements</u>. Together, these standards and essential elements identify the most important ideas in the study of geography. These ideas are expressed in two lists.

The first list is the national geography standards. This is a list of 18 basic ideas that are central to the study of geography. These standards are listed in black type on the chart below.

The essential elements are based on the geography standards. Each element is a big idea that links several standards together. The six essential elements are The World in Spatial Terms, Places and Regions, Physical Systems, Human Systems, Environment and Society, and The Uses of Geography. On the chart, they are shown in purple.

Read through that list again. Do you see any similarities between geography's six essential elements and its five themes? You probably do. The two systems are very similar because the six essential elements build on the five themes.

ACADEMIC VOCABULARY
element part

The Essential Elements and Geography Standards

The World in Spatial Terms

- How to use maps and other geographic representations, tools, and technologies to acquire, process, and report information from a spatial perspective
- How to use mental maps to organize information about people, places, and environments in a spatial context
- How to analyze the spatial organization of people, places, and environments on Earth's surface

Places and Regions

- The physical and human characteristics of places
- How people create regions to interpret Earth's complexity
- How culture and experience influence people's perceptions of places and regions

Physical Systems

- The physical processes that shape the patterns of Earth's surface
- The characteristics and spatial distribution of ecosystems on Earth's surface

Human Systems

- The characteristics, distributions, and migration of human populations on Earth's surface
- The characteristics, distribution, and complexity of Earth's cultural mosaics
- The patterns and networks of economic interdependence on Earth's surface
- The processes, patterns, and functions of human settlement
- How the forces of cooperation and conflict among people influence the division and control of Earth's surface

Environment and Society

- How human actions modify the physical environment
- How physical systems affect human systems
- Changes that occur in the meaning, use, distribution, and importance of resources

The Uses of Geography

- How to apply geography to interpret the past
- How to apply geography to interpret the present and plan for the future

BOOK
Geography for Life

The six essential elements were first outlined in a book called Geography for Life. *In that book, the authors—a diverse group of geographers and teachers from around the United States—explained why the study of geography is important.*

❝Geography *is* for life in every sense of that expression: lifelong, life-sustaining, and life-enhancing. Geography is a field of study that enables us to find answers to questions about the world around us—about where things are and how and why they got there.❞

❝Geography focuses attention on exciting and interesting things, on fascinating people and places, on things worth knowing because they are absorbing and because knowing about them lets humans make better-informed and, therefore, wiser decisions.❞

❝With a strong grasp of geography, people are better equipped to solve issues at not only the local level but also the global level.❞

–from *Geography for Life,*
by the Geography Education Standards Project

ANALYSIS SKILL **ANALYZING PRIMARY SOURCES**

Why do the authors of these passages think that people should study geography?

For example, the element Places and Regions combines two of the five themes of geography—Place and Regions. Also, the element called Environment and Society deals with many of the same issues as the theme Human-Environment Interaction.

There are also some basic differences between the essential elements and the themes. For example, the last element, The Uses of Geography, deals with issues not covered in the five themes. This element examines how people can use geography to plan the landscapes in which they live.

Throughout this book, you will notice references to both the themes and the essential elements. As you read, use these themes and elements to help you organize your own study of geography.

READING CHECK **Summarizing** What are the six essential elements of geography?

SUMMARY AND PREVIEW You have just learned about the themes and elements of geography. Next, you will explore the branches into which the field is divided.

---

go.hrw.com
Online Quiz
KEYWORD: SGA7 HP1

Section 2 Assessment

Reviewing Ideas, Terms, and Places

1. **a. Define** What is the difference between a place's **absolute location** and its **relative location**? Give one example of each type of location.
   b. Contrast How are the themes of Location and Place different?
   c. Elaborate How does using the five themes help geographers understand the places they study?
2. **a. Identify** Which of the five themes of geography is associated with airports, highways, and the migration of people from one place to another?
   b. Explain How are the geography standards and the six essential elements related?
   c. Compare How are the six essential elements similar to the five themes of geography?

Critical Thinking

3. **Categorizing** Draw a chart like the one below. Use your notes to list the five themes of geography, explain each of the themes, and list one feature of your city or town that relates to each.

Theme				
Explanation				
Feature				

FOCUS ON WRITING

4. **Including Themes and Essential Elements** The five themes and six essential elements are central to a geographer's job. How will you mention them in your job description? Write down some ideas.

Social Studies Skills

Chart and Graph

Critical Thinking

Geography

Study

Analyzing Satellite Images

Learn

In addition to maps and globes, satellite images are among the geographer's most valuable tools. Geographers use two basic types of these images. The first type is called true color. These images are like photographs taken from high above Earth's surface. The colors in these images are similar to what you would see from the ground. Vegetation, for example, appears green.

The other type of satellite image is called an infrared image. Infrared images are taken using a special type of light. These images are based on heat patterns, and so the colors on them are not what we might expect. Bodies of water appear black, for example, since they give off little heat.

Practice

Use the satellite images on this page to answer the following questions.

1. On which image is vegetation red?

2. Which image do you think probably looks more like Italy does from the ground?

True color satellite image of Italy

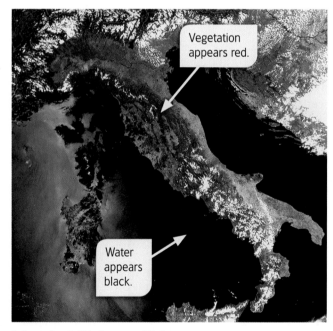

Infrared satellite image of Italy

Apply

Search the Internet to find a satellite image of your state or region. Determine whether the image is true color or infrared. Then write three statements that describe what you see on the image.

The Branches of Geography

What You Will Learn...

Main Ideas

1. Physical geography is the study of landforms, water bodies, and other physical features.
2. Human geography focuses on people, their cultures, and the landscapes they create.
3. Other branches of geography examine specific aspects of the physical or human world.

The Big Idea

Geography is divided into two main branches—physical and human geography.

Key Terms

physical geography, *p. 16*
human geography, *p. 18*
cartography, *p. 19*
meteorology, *p. 20*

 TAKING NOTES Draw two large circles like the ones below in your notebook. As you read this section, take notes about one of the main branches of geography in each circle.

If YOU lived there...

You are talking to two friends about the vacations their families will take this summer. One friend says that his family is going to the Grand Canyon. He is very excited about seeing the spectacular landscapes in and around the canyon. Your other friend's family is going to visit Nashville, Tennessee. She is looking forward to trying new foods at the city's restaurants and touring its museums.

Which vacation sounds more interesting? Why?

BUILDING BACKGROUND Geography is the study of the world and its many features. Some of those features are physical, like the Grand Canyon. Others are human, like food and culture. The main branches of geography focus on the study of these types of features.

Physical Geography

Think about a jigsaw puzzle. Seen as a whole, the puzzle shows a pretty or interesting picture. To see that picture, though, you have to put all the puzzle pieces together. Before you assemble them, the pieces do not give you a clear idea of what the puzzle will look like when it is assembled. After all, each piece contains only a tiny portion of the overall image.

In many ways, geography is like a huge puzzle. It is made up of many branches, or divisions. Each of these branches focuses on a single part of the world. Viewed separately, none of these branches shows us the whole world. Together, however, the many branches of geography improve our understanding of our planet and its people.

Geography's two main branches are physical geography and human geography. The first branch, **physical geography**, is the study of the world's physical features—its landforms, bodies of water, climates, soils, and plants. Every place in the world has its own unique combination of these features.

Physical Geography

The study of Earth's physical features, including rivers, mountains, oceans, weather, and other features, such as Victoria Falls in southern Africa

Human Geography

The study of Earth's people, including their ways of life, homes, cities, beliefs, and customs, like those of these children in Malawi, a country in central Africa

Geography

The study of Earth's physical and cultural features

The Physical World

What does it mean to say that physical geography is the study of physical features? Physical geographers want to know all about the different features found on our planet. They want to know where plains and mountain ranges are, how rivers flow across the landscape, and why different amounts of rain fall from place to place.

More importantly, however, physical geographers want to know what causes the different shapes on Earth. They want to know why mountain ranges rise up where they do and what causes rivers to flow in certain directions. They also want to know why various parts of the world have very different weather and climate patterns.

To answer these questions, physical geographers take detailed measurements. They study the heights of mountains and the temperatures of places. To track any changes that occur over time, physical geographers keep careful records of all the information they collect.

Uses of Physical Geography

Earth is made up of hundreds of types of physical features. Without a complete understanding of what these features are and the effect they have on the world's people and landscapes, we cannot fully understand our world. This is the major reason that geographers study the physical world—to learn how it works.

There are also other, more specific reasons for studying physical geography, though. Studying the changes that take place on our planet can help us prepare to live with those changes. For example, knowing what causes volcanoes to erupt can help us predict eruptions. Knowing what causes terrible storms can help us prepare for them. In this way, the work of physical geographers helps us adjust to the dangers and changes of our world.

READING CHECK **Analyzing** What are some features in your area that a physical geographer might study?

BIOGRAPHY

Eratosthenes

(c. 276–c. 194 BC)

Did you know that geography is over 2,000 years old? Actually, the study of the world is even older than that, but the first person ever to use the word *geography* lived then. His name was Eratosthenes (er-uh-TAHS-thuh-neez), and he was a Greek scientist and librarian. With no modern instruments of any kind, Eratosthenes figured out how large Earth is. He also drew a map that showed all of the lands that the Greeks knew about. Because of his many contributions to the field, Eratosthenes has been called the Father of Geography.

Generalizing Why is Eratosthenes called the Father of Geography?

Human Geography

The physical world is only one part of the puzzle of geography. People are also part of the world. **Human geography** is the study of the world's people, communities, and landscapes. It is the second major branch of geography.

The Human World

Put simply, human geographers study the world's people, past and present. They look at where people live and why. They ask why some parts of the world have more people than others, and why some places have almost no people at all.

Human geographers also study what people do. What jobs do people have? What crops do they grow? What makes them move from place to place? These are the types of questions that geographers ask about people around the world.

Because people's lives are so different around the world, no one can study every aspect of human geography. As a result, human geographers often specialize in a smaller area of study. Some may choose to study only the people and landscapes in a certain region. For example, a geographer may study only the lives of people who live in Africa.

Other geographers choose not to limit their studies to one place. Instead, they may choose to examine only one aspect of people's lives. For example, a geographer could study only economics, politics, or city life. However, that geographer may compare economic patterns in various parts of the world to see how they differ.

Uses of Human Geography

Although every culture is different, people around the world have some common needs. All people need food and water. All people need shelter. All people need to deal with other people in order to survive.

Human geographers study how people in various places address their needs. They look at the foods people eat and the types of governments they form. The knowledge they gather can help us better understand people in other cultures. Sometimes this type of understanding can help people improve their landscapes and situations.

On a smaller scale, human geographers can help people design their cities and towns. By understanding where people go and what they need, geographers can help city planners place roads, shopping malls, and schools. Geographers also study the effect people have on the world. As a result, they often work with private groups and government agencies who want to protect the environment.

READING CHECK **Summarizing** What do human geographers study?

Other Fields of Geography

Physical geography and human geography are the two largest branches of the subject, but they are not the only ones. Many other fields of geography exist, each one devoted to studying one aspect of the world.

Most of these fields are smaller, more specialized areas of either physical or human geography. For example, economic geography—the study of how people make and spend money—is a branch of human geography. Another specialized branch of human geography is urban geography, the study of cities and how people live in them. Physical geography also includes many fields, such as the study of climates. Other fields of physical geography are the studies of soils and plants.

Cartography

One key field of geography is **cartography**, the science of making maps. You have already seen how important maps are to the study of geography. Without maps, geographers would not be able to study where things are in the world.

In the past, maps were always drawn by hand. Many were not very accurate. Today, though, most maps are made using computers and satellite images. Through advances in mapmaking, we can make accurate maps on almost any scale, from the whole world to a single neighborhood, and keep them up to date. These maps are not only used by geographers. For example, road maps are used by people who are planning long trips.

CONNECTING TO Technology

Computer Mapping

In the past, maps were drawn by hand. Making a map was a slow process. Even the simplest map took a long time to make. Today, however, cartographers have access to tools people in the past—even people who lived just 50 years ago—never imagined. The most important of these tools are computers.

Computers allow us to make maps quickly and easily. In addition, they let us make new types of maps that people could not make in the past.

The map shown here, for example, was drawn on a computer. It shows the number of computer users in the United States who were connected to the Internet on a particular day. Each of the lines that rises off of the map represents a city in which people were using the Internet. The color of the line indicates the number of computer users in that city. As you can see, this data resulted in a very complex map.

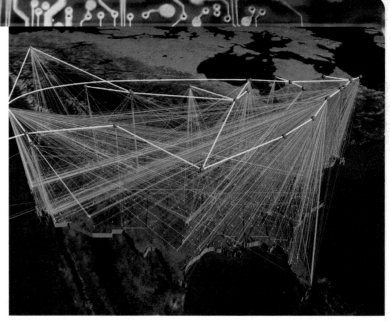

Making such a map required cartographers to sort through huge amounts of complex data. Such sorting would not have been possible without computers.

Contrasting How are today's maps different from those created in the past?

Meteorology is the study of weather. This meteorologist is using computers to follow and predict the movement of a powerful storm.

Meteorology

Have you ever seen the weather report on television? If so, you have seen the results of another branch of geography. This branch is called **meteorology**, the study of weather and what causes it.

Meteorologists study weather patterns in a particular area. Then they use the information to predict what the weather will be like in the coming days. Their work helps people plan what to wear and what to do on any given day. At the same time, their work can save lives by predicting the arrival of terrible storms. These predictions are among the most visible ways in which the work of geographers affects our lives every day.

READING CHECK **Finding Main Ideas** What are some major branches of geography?

SUMMARY AND PREVIEW In this section, you learned about two main branches of geography, physical and human. In the next chapter, you will learn more about the physical features that surround us and the processes that create them.

Hydrology

FOCUS ON READING
What do you already know about drinking water?

Another important branch of geography is hydrology, the study of water on Earth. Geographers in this field study the world's river systems and rainfall patterns. They study what causes floods and how people in cities can get safe drinking water. They also work to measure and protect the world's supply of water.

Section 3 Assessment

go.hrw.com
Online Quiz
KEYWORD: SGA7 HP1

Reviewing Ideas, Terms, and Places

1. a. **Define** What is **physical geography**?
   b. **Explain** Why do we study physical geography?
2. a. **Identify** What are some things that people study as part of **human geography**?
   b. **Summarize** What are some ways in which the study of human geography can influence our lives?
   c. **Evaluate** Which do you think would be more interesting to study, physical geography or human geography? Why?
3. a. **Identify** What are two specialized fields of geography?
   b. **Analyze** How do cartographers contribute to the work of other geographers?

Critical Thinking

4. **Comparing and Contrasting** Draw a diagram like the one shown here. In the left circle, list three features of physical geography from your notes. In the right circle, list three features of human geography. Where the circles overlap, list one feature they share.

   Physical   Human

FOCUS ON WRITING

5. **Choosing a Branch** Your job description should point out to people that there are many branches of geography. How will you note that?

Chapter Review

Geography's Impact
video series
Review the video to answer the closing question:
Why do you think it might be valuable to know the absolute location of a place?

Visual Summary

Use the visual summary below to help you review the main ideas of the chapter.

QUICK FACTS

Physical geography—the study of the world's physical features—is one main branch of geography.

Human geography—the study of the world's people and how they live—is the second main branch.

Geographers use many tools to study the world. The most valuable of these tools are maps.

Reviewing Vocabulary, Terms, and Places

Match the words in the columns with the correct definitions listed below.

1. geography
2. physical geography
3. human geography
4. element
5. meteorology

6. region
7. cartography
8. map
9. landscape
10. globe

a. a part of the world that has one or more common features that make it different from surrounding areas

b. a flat drawing of part of Earth's surface

c. a part

d. a spherical model of the planet

e. the study of the world's physical features

f. the study of weather and what causes it

g. the study of the world, its people, and the landscapes they create

h. the science of making maps

i. the physical and human features that define an area and make it different from other places

j. the study of people and communities

Comprehension and Critical Thinking

SECTION 1 *(Pages 4–9)*

11. a. Identify What are three levels at which a geographer might study the world? Which of these levels covers the largest area?

b. Compare and Contrast How are maps and globes similar? How are they different?

c. Elaborate How might satellite images and computers help geographers improve their knowledge of the world?

SECTION 2 *(Pages 10–14)*

12. a. Define What do geographers mean when they discuss an area's landscape?

b. Explain Why did geographers create the five themes and the six essential elements?

c. Predict How might the five themes and six essential elements help you in your study of geography?

SECTION 3 *(Pages 16–20)*

13. a. Identify What are the two main branches of geography? What does each include?

b. Summarize How can physical geography help people adjust to the dangers of the world?

c. Elaborate Why do geographers study both physical and human characteristics of places?

Using the Internet

go.hrw.com
KEYWORD: SGA7 CH1

14. Activity: Using Maps What does your town or community look like? What can be found there? Maps can help you understand your community and learn about its features. Enter the activity keyword to learn more about maps and how they can help you better understand your community. Then search the Internet to find a map of your community. Use the map to find the locations of at least five important features. For example, you might locate your school, the library, a park, or major highways. Be creative and find other places you think your classmates should be aware of.

Social Studies Skills

Analyzing Satellite Images *Use the satellite images of Italy from the Social Studies Skills lesson in this chapter to answer the following questions.*

15. On which image do forests appear more clearly, the true-color or the infrared image?

16. What color do you think represents mountains on the infrared satellite image?

17. Why might geographers use satellite images like these while making maps of Italy?

18. Using Prior Knowledge Create a chart with three columns. In the first column list what you knew before you read the chapter. In the second column list what you learned in the chapter. In the third column list questions that you now have about geography.

19. Writing Your Job Description Review your notes on the job of a geographer. Then write your job description. You should begin your description by explaining why the job is important. Then identify the job's tasks and responsibilities. Finally, tell what kind of person might do well as a geographer.

Map Activity

20. Sketch Map Draw a map that shows your school and the surrounding neighborhood. Your map does not have to be complicated, but you should include major features like streets and buildings. Use the map shown here as an example.

DIRECTIONS: Read questions 1 through 7 and write the letter of the best response. Then read question 8 and write your own well-constructed response.

1 Which of the following subjects would a human geographer study the most?

A mountains

B populations

C rivers

D volcanoes

2 The study of weather is called

A meteorology.

B hydrology.

C social science.

D cartography.

3 A region is an area that has

A one or more common features.

B no people living in it.

C few physical features.

D set physical boundaries.

4 How many essential elements of geography have geographers identified?

A two

B four

C six

D eight

5 The physical and human characteristics that define an area are its

A landscape.

B location.

C region.

D science.

The United States

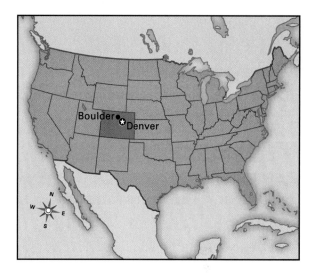

6 Which of the five themes of geography would a geographer most likely study using this map?

A movement

B location

C human-environment interaction

D landscape

7 The smallest level at which a geographer might study a place is

A microscopic.

B local.

C regional.

D global.

8 **Extended Response** Look at the map of the United States above. Do you think this map is more likely to be used by a physical geographer or by a human geographer? Give two reasons for your answer. Then write two statements about what a geographer could find on this map.

CHAPTER 2
Planet Earth

What You Will Learn...

In this chapter you will learn about important processes on planet Earth. You will discover how Earth's movements affect the energy we receive from the sun, how water affects life, and how Earth's landforms were made.

SECTION 1
Earth and the Sun's Energy **26**

SECTION 2
Water on Earth **30**

SECTION 3
The Land **35**

FOCUS ON READING AND WRITING

Using Word Parts Sometimes you can figure out the meaning of a word by looking at its parts. A root is the base of the word. A prefix attaches to the beginning, and a suffix attaches to the ending. When you come across a word you don't know, check to see whether you recognize its parts. **See the lesson, Using Word Parts, on page 107.**

Writing a Haiku Join the poets who have celebrated our planet for centuries. Write a haiku, a short poem, about planet Earth. As you read the chapter, gather information about changes in the sun's energy, Earth's water supply, and shapes on the land. Then choose the most intriguing information to include in your haiku.

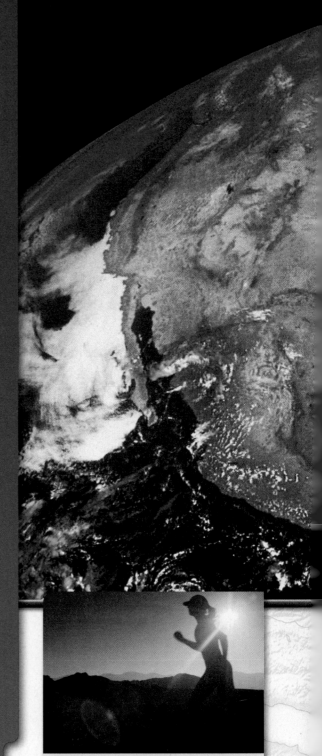

Energy from the Sun The planet's movement creates differences in the amount of energy Earth receives from the sun.

Many of Earth's features are visible from space. This photo, taken from a satellite orbiting the planet, shows part of the North American continent.

Which of Earth's features are visible in this photo?

HOLT

Geography's Impact
video series
Watch the video to understand the impact of water on Earth.

Land Forces on and under Earth's surface have shaped the different landforms on our planet. Geographers study how mountains and other land-forms were made.

Water on Earth Water is essential for life on Earth. Much of the planet's water supply is stored in Earth's oceans and ice caps.

25

Earth and the Sun's Energy

What You Will Learn...

Main Ideas

1. Earth's movement affects the amount of energy we receive from the sun.
2. Earth's seasons are caused by the planet's tilt.

The Big Idea

Earth's movement and the sun's energy interact to create day and night, temperature changes, and the seasons.

Key Terms

solar energy, *p. 26*
rotation, *p. 26*
revolution, *p. 27*
latitude, *p. 27*
tropics, *p. 29*

TAKING NOTES As you read, take notes on Earth's movement and the seasons. Use a chart like the one below to organize your notes.

Earth's Movement	The Seasons

If YOU lived there...

You live in Chicago and have just won an exciting prize—a trip to Australia during winter vacation in January. As you prepare for the trip, your mother reminds you to pack shorts and a swimsuit. You are confused. In January you usually wear winter sweaters and a heavy jacket.

Why is the weather so different in Australia?

BUILDING BACKGROUND Seasonal differences in weather are an important result of Earth's constant movement. As the planet moves, we experience changes in the amount of energy we receive from the sun. Geographers study and explain why different places on Earth receive differing amounts of energy from the sun.

Earth's Movement

Energy from the sun helps crops grow, provides light, and warms Earth. It even influences the clothes we wear, the foods we eat, and the sports we play. All life on Earth requires **solar energy**, or energy from the sun, to survive. The amount of solar energy places on Earth receive changes constantly. Earth's rotation, revolution, and tilt, as well as latitude, all affect the amount of solar energy parts of the planet receive from the sun.

Rotation

Imagine that Earth has a rod running through it from the North Pole to the South Pole. This rod represents Earth's axis—an imaginary line around which a planet turns. As Earth spins on its axis, different parts of the planet face the sun. It takes Earth 24 hours, or one day, to complete this rotation. A **rotation** is one complete spin of Earth on its axis. As Earth rotates during this 24-hour period, it appears to us that the sun moves across the sky. The sun seems to rise in the east and set in the west. The

Solar Energy

Earth's tilt and rotation cause changes in the amount of energy we receive from the sun. As Earth rotates on its axis, energy from the sun creates periods of day and night. Earth's tilt causes some locations, especially those close to the equator, to receive more direct solar energy than others.

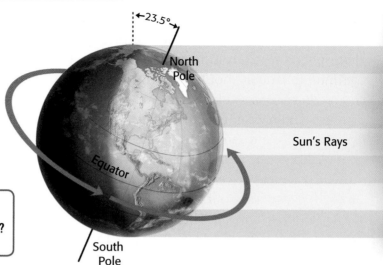

23.5°

North Pole

Sun's Rays

Equator

South Pole

ANALYSIS SKILL **ANALYZING VISUALS**

Is the region north or south of the equator receiving more solar energy? How can you tell?

sun, however, does not move. It is actually Earth's rotation that creates the sense of the sun's movement.

Earth's rotation also explains why day changes to night. As you can see in the illustration, solar energy strikes only the half of Earth facing the sun. Warmth and light from the sun create daytime. At the same time, the half of the planet facing away from the sun experiences the cooler temperatures and darkness of night. Earth's rotation causes regular shifts from day to night. As a result, levels of solar energy on Earth constantly change.

Revolution

As Earth spins on its axis, it also follows a path, or orbit, around the sun. Earth's orbit around the sun is not a perfect circle. Sometimes the orbit takes Earth closer to the sun, and at other times the orbit takes it farther away. It takes 365¼ days for Earth to complete one **revolution**, or trip around the sun. We base our calendar year on the time it takes Earth to complete its orbit around the sun. To allow for the fraction of a day, we add an extra day—February 29—to our calendar every four years.

Tilt and Latitude

Another **factor** affecting the amount of solar energy we receive is the planet's tilt. As the illustration shows, Earth's axis is not straight up and down. It is actually tilted at an angle of 23½ degrees from vertical. At any given time of year, some locations on Earth are tilting away from the sun, and others are tilting toward it. Places tilting toward the sun receive more solar energy and experience warmer temperatures. Those tilting away from the sun receive less solar energy and experience cooler temperatures.

A location's **latitude**, the distance north or south of Earth's equator, also affects the amount of solar energy it receives. Low-latitude areas, those near the equator like Hawaii, receive direct rays from the sun all year. These direct rays are more intense and produce warmer temperatures. Regions with high latitudes, like Antarctica, are farther from the equator. As a result, they receive indirect rays from the sun and have colder temperatures.

ACADEMIC VOCABULARY

factor cause

READING CHECK **Finding Main Ideas** What factors affect the solar energy Earth receives?

The Seasons

Does the thought of snow in July or 100-degree temperatures in January seem odd to you? It might if you live in the Northern Hemisphere, where cold temperatures are common in January, not July. The planet's changing seasons explain why we often connect certain weather with specific times of the year, like snow in January. Seasons are periods during the year that are known for a particular type of weather. Many places on Earth experience four seasons—winter, spring, summer, and fall. These seasons are based on temperature and length of day. In some parts of the world, however, seasons are based on the amount of rainfall.

FOCUS ON READING
The prefix *hemi*- means half. What does the word *hemisphere* mean?

Winter and Summer

The change in seasons is created by Earth's tilt. As you can see in the illustration below, while one of Earth's poles tilts away from the sun, the other tilts toward it. During winter part of Earth is tilted away from the sun, causing less direct solar energy, cool temperatures, and less daylight. Summer occurs when part of Earth is tilted toward the sun. This creates more direct solar energy, warmer temperatures, and longer periods of daylight.

Because of Earth's tilt, the Northern and Southern hemispheres experience opposite seasons. As the North Pole tilts toward the sun in summer, the South Pole tilts away

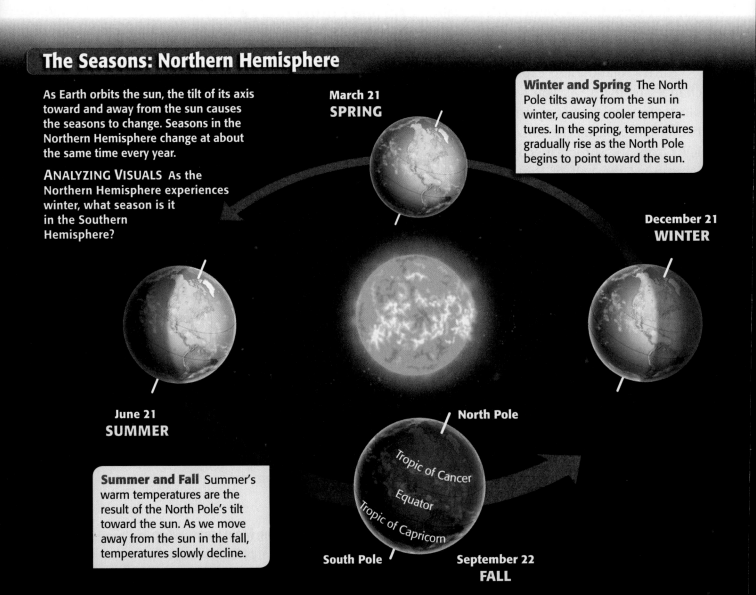

The Seasons: Northern Hemisphere

As Earth orbits the sun, the tilt of its axis toward and away from the sun causes the seasons to change. Seasons in the Northern Hemisphere change at about the same time every year.

ANALYZING VISUALS As the Northern Hemisphere experiences winter, what season is it in the Southern Hemisphere?

March 21 SPRING

Winter and Spring The North Pole tilts away from the sun in winter, causing cooler temperatures. In the spring, temperatures gradually rise as the North Pole begins to point toward the sun.

December 21 WINTER

June 21 SUMMER

Summer and Fall Summer's warm temperatures are the result of the North Pole's tilt toward the sun. As we move away from the sun in the fall, temperatures slowly decline.

North Pole
Tropic of Cancer
Equator
Tropic of Capricorn
South Pole

September 22 FALL

from it. As a result, the Southern Hemisphere experiences winter. Likewise, when it is spring in the Northern Hemisphere, it is fall in the Southern Hemisphere.

Spring and Fall

As Earth orbits the sun, there are periods when the poles tilt neither toward nor away from the sun. These periods mark spring and fall. During the spring, as part of Earth begins to tilt toward the sun, solar energy increases. Temperatures slowly start to rise, and days grow longer. In the fall the opposite occurs as winter approaches. Solar energy begins to decrease, causing cooler temperatures and shorter days.

Rainfall and Seasons

Some regions on Earth have seasons marked by rainfall rather than temperature. This is true in the **tropics**, regions close to the equator. At certain times of year, winds bring either dry or moist air to the tropics, creating wet and dry seasons. In India, for example, seasonal winds called monsoons bring heavy rains from June to October and dry air from November to January.

READING CHECK **Identifying Cause and Effect** What causes the seasons to change?

The Midnight Sun

Can you imagine going to sleep late at night with the sun shining in the sky? People who live near the Arctic and Antarctic circles experience this every summer, when they can receive up to 24 hours of sunlight a day. The time-lapse photo below shows a typical sunset during this period—except the sun never really sets! This phenomenon is known as the midnight sun. For locations like Tromso, Norway, this means up to two months of constant daylight each summer. People living near Earth's poles often use the long daylight hours to work on outdoor projects in preparation for winter, when they can receive 24 hours of darkness a day.

Predicting How might people's daily lives be affected by the midnight sun?

SUMMARY AND PREVIEW Solar energy is crucial for all life on the planet. Earth's position and movements affect the amount of energy we receive from the sun and determine our seasons. Next, you will learn about Earth's water supply and its importance to us.

Section 1 Assessment

Reviewing Key Ideas, Terms, and Places

1. **a. Identify** What is **solar energy**, and how does it affect Earth?
   b. Analyze How do **rotation** and tilt each affect the amount of solar energy that different parts of Earth receive?
   c. Predict What might happen if Earth received less solar energy than it currently does?
2. **a. Describe** Name and describe Earth's seasons.
   b. Contrast How are seasons different in the Northern and Southern hemispheres?
   c. Elaborate How might the seasons affect human activities?

Critical Thinking

3. **Identifying Cause and Effect** Use your notes and the diagram to identify the causes of seasons.

Cause	
---	Effect: Earth's changing seasons
Cause	

FOCUS ON WRITING

4. **Describing the Seasons** What are the seasons like where you live? In your notebook, jot down a few notes that describe the changing seasons.

Water on Earth

What You Will Learn...

Main Ideas

1. Salt water and freshwater make up Earth's water supply.
2. In the water cycle, water circulates from Earth's surface to the atmosphere and back again.
3. Water plays an important role in people's lives.

The Big Idea

Water is a dominant feature on Earth's surface and is essential for life.

Key Terms

freshwater, *p. 31*
glaciers, *p. 31*
surface water, *p. 31*
precipitation, *p. 31*
groundwater, *p. 32*
water vapor, *p. 32*
water cycle, *p. 33*

TAKING NOTES As you read, take notes about Earth's water, the water cycle, and how water affects our lives. Use a diagram like the one below to organize your notes.

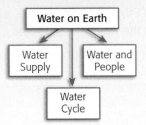

If **YOU** lived there...

You live in the desert Southwest, where heavy water use and a lack of rainfall have led to water shortages. Your city plans to begin a water conservation program that asks people to limit how much water they use. Many of your neighbors have complained that the program is unnecessary. Others support the plan to save water.

How do you feel about the city's water plan?

BUILDING BACKGROUND Although water covers much of Earth's surface, water shortages, like those in the American Southwest, are common all over the planet. Because water is vital to the survival of all living things, geographers study Earth's water supply.

Earth's Water Supply

Think of the different uses for water. We use water to cook and clean, we drink it, and we grow crops with it. Water is used for recreation, to generate electricity, and even to travel from place to place. Water is perhaps the most important and abundant resource on Earth. In fact, water covers some two-thirds of the planet. Understanding Earth's water supply and how it affects our lives is an important part of geography.

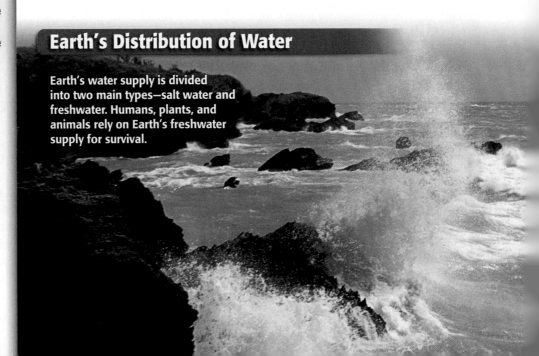

Earth's Distribution of Water

Earth's water supply is divided into two main types—salt water and freshwater. Humans, plants, and animals rely on Earth's freshwater supply for survival.

Salt Water

Although water covers much of the planet, we cannot use most of it. About 97 percent of the Earth's water is salt water. Because salt water contains high levels of salt and other minerals, it is unsafe to drink.

In general, salt water is found in Earth's oceans. Oceans are vast bodies of water covering some 71 percent of the planet's surface. Earth's oceans are made up of smaller bodies of water such as seas, gulfs, bays, and straits. Altogether, Earth's oceans cover some 139 million square miles (360 million sq km) of the planet's surface.

Some of Earth's lakes contain salt water. The Great Salt Lake in Utah, for example, is a saltwater lake. As salt and other minerals have collected in the lake, which has no outlet, the water has become salty.

Freshwater

Since the water in Earth's oceans is too salty to use, we must rely on other sources for freshwater. **Freshwater**, or water without salt, makes up only about 3 percent of our total water supply. Much of that freshwater is locked in Earth's **glaciers**, large areas of slow-moving ice, and in the ice of the Antarctic and Arctic regions. Most of the freshwater we use everyday is found in lakes, in rivers, and under Earth's surface.

One form of freshwater is surface water. **Surface water** is water that is found in Earth's streams, rivers, and lakes. It may seem that there is a great deal of water in our lakes and rivers, but only a tiny amount of Earth's water supply—less than 1 percent—comes from surface water.

Streams and rivers are a common source of surface water. Streams form when precipitation collects in a narrow channel and flows toward the ocean. **Precipitation** is water that falls to Earth's surface as rain, snow, sleet, or hail. In turn, streams join together to form rivers. Any smaller stream or river that flows into a larger stream or river is called a tributary. For example, the Missouri River is the largest tributary of the Mississippi River.

Lakes are another important source of surface water. Some lakes were formed as rivers filled low-lying areas with water. Other lakes, like the Great Lakes along the U.S.–Canada border, were formed when glaciers carved deep holes in Earth's surface and deposited water as they melted.

Most of Earth's available freshwater is stored underground. As precipitation falls to Earth, much of it is absorbed into the ground, filling spaces in the soil and rock.

Salt Water Earth's oceans contain some 97 percent of the planet's water supply. Unfortunately, this water is too salty to drink.

Freshwater Freshwater from lakes, rivers, and streams makes up only a fraction of Earth's water supply.

Water found below Earth's surface is called **groundwater**. In some places on Earth, groundwater naturally bubbles from the ground as a spring. More often, however, people obtain groundwater by digging wells, or deep holes dug into the ground to reach the water.

READING CHECK **Contrasting** How is salt water different from freshwater?

**✱Interactive
Close-up**

The Water Cycle

Energy from the sun drives the water cycle. Surface water evaporates into Earth's atmosphere, where it condenses, then falls back to Earth as precipitation. This cycle repeats continuously, providing us with a fairly constant water supply.

go.hrw.com (KEYWORD: SGA7 CH2)

The Water Cycle

When you think of water, you probably visualize a liquid—a flowing stream, a glass of ice-cold water, or a wave hitting the beach. But did you know that water is the only substance on Earth that occurs naturally as a solid, a liquid, and a gas? We see water as a solid in snow and ice and as a liquid in oceans and rivers. Water also occurs in the air as an invisible gas called **water vapor**.

Water is always moving. As water heats up and cools down, it moves from the planet's surface to the atmosphere, or the mass of air that surrounds Earth. One of the most important processes in nature

Condensation occurs when water vapor cools and forms clouds.

When the droplets in clouds become too heavy, they fall to Earth as precipitation.

Runoff is excess precipitation that flows over land into rivers, streams, and oceans.

ANALYSIS SKILL **ANALYZING VISUALS**
How does evaporation differ from precipitation?

is the water cycle. The **water cycle** is the movement of water from Earth's surface to the atmosphere and back.

The sun's energy drives the water cycle. As the sun heats water on Earth's surface, some of that water evaporates, or turns from liquid to gas, or water vapor. Water vapor then rises into the air. As the vapor rises, it cools. The cooling causes the water vapor to condense, or change from a vapor into tiny liquid droplets. These droplets join together to form clouds. If the droplets become heavy enough, precipitation occurs—that is, the water falls back to Earth as rain, snow, sleet, or hail.

When that precipitation falls back to Earth's surface, some of the water is absorbed into the soil as groundwater. Excess water, called runoff, flows over land and collects in streams, rivers, and oceans. Because the water cycle is constantly repeating, it allows us to maintain a fairly constant supply of water on Earth.

READING CHECK **Finding Main Ideas** What is the water cycle?

As energy from the sun heats water on Earth's surface, the water evaporates, or turns to water vapor, and rises to the atmosphere.

Water and People

How many times a day do you think about water? Many of us rarely give it a second thought, yet water is crucial for survival. Water problems such as the lack of water, polluted water, and flooding are concerns for people all around the world. Water also provides us with <u>countless</u> benefits, such as energy and recreation.

Water Problems

One of the greatest water problems people face is a lack of available freshwater. Many places face water shortages as a result of droughts, or long periods of lower-than-normal precipitation. Another cause of water shortages is overuse. In places like the southwestern United States, where the population has grown rapidly, the heavy demand for water has led to shortages.

Even where water is plentiful, it may not be clean enough to use. If chemicals and household wastes make their way into streams and rivers, they can contaminate the water supply. Polluted water can carry diseases. These diseases may harm humans, plants, and animals.

Flooding is another water problem that affects people around the world. Heavy rains often lead to flooding, which can damage property and threaten lives. One example of dangerous flooding occurred in Bangladesh in 2004. Floods there destroyed roads and schools and left some 25 million people homeless.

Water's Benefits

Water does more than just quench our thirst. It provides us with many benefits, such as food, power, and even recreation.

Water's most important benefit is that it provides us with food to eat. Everything we eat depends on water. For example, fruits and vegetables need water to grow.

FOCUS ON READING
Look at the word *countless* in this paragraph. The suffix *-less* means unable to. What does *countless* mean?

The Benefits of Water

Many people take advantage of the recreational and agricultural benefits that water provides.

Animals also need water to live and grow. As a result, we use water to farm and raise animals so that we will have food to eat.

Water is also an important source of energy. Using dams, we harness the power of moving water to produce electricity. Electricity provides power to air-condition or heat our homes, to run our washers and dryers, and to keep our food cold.

Water also provides us with recreation. Rivers, lakes, and oceans make it possible for us to swim, to fish, to surf, or to sail a boat. Although recreation is not critical for our survival, it does make our lives richer and more enjoyable.

READING CHECK Summarizing How does water affect people's lives?

SUMMARY AND PREVIEW In this section you learned that water is essential for life on Earth. Next, you will learn about the shapes on Earth's surface.

---

go.hrw.com
Online Quiz
KEYWORD: SGA7 HP2

Section 2 Assessment

Reviewing Key Ideas, Terms, and Places

1. **a. Describe** Name and describe the different types of water that make up Earth's water supply.
   b. Analyze Why is only a small percentage of Earth's **freshwater** available to us?
   c. Elaborate In your opinion, which is more important—**surface water** or **groundwater**? Why?
2. **a. Recall** What drives the **water cycle**?
   b. Make Inferences From what bodies of water do you think most evaporation occurs? Why?
3. **a. Define** What is a drought?
   b. Analyze How does water support life on Earth?
   c. Evaluate What water problem do you think is most critical in your community? Why?

Critical Thinking

4. **Sequencing** Draw the graphic organizer at right. Then use your notes and the graphic organizer to identify the stages in Earth's water cycle.

FOCUS ON WRITING

5. **Learning about Water** Consider what you have learned about water in this section. How might you describe water in your haiku? What words might you use to describe Earth's water supply?

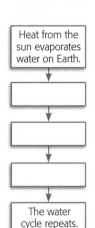

Heat from the sun evaporates water on Earth.

↓

↓

↓

The water cycle repeats.

The Land

If YOU lived there...

You live in the state of Washington. All your life, you have looked out at the beautiful, cone-shaped peaks of nearby mountains. One of them is Mount Saint Helens, an active volcano. You know that in 1980 it erupted violently, blowing a hole in the mountain and throwing ash and rock into the sky. Since then, scientists have watched the mountain carefully.

How do you feel about living near a volcano?

BUILDING BACKGROUND Over billions of years, many different forces have changed Earth's surface. Processes deep underground have built up landforms and even shifted the position of continents. Wind, water, and ice have also shaped the planet's landforms. Changes in Earth's surface continue to take place.

Landforms

Do you know the difference between a valley and a volcano? Can you tell a peninsula from a plateau? If you answered yes, then you are familiar with some of Earth's many landforms. **Landforms** are shapes on the planet's surface, such as hills or mountains. Landforms make up the landscapes that surround us, whether it's the rugged mountains of central Colorado or the flat plains of Oklahoma.

Earth's surface is covered with landforms of many different shapes and sizes. Some important landforms include:

- mountains, land that rises higher than 2,000 feet (610 m)
- valleys, areas of low land located between mountains or hills
- plains, stretches of mostly flat land
- islands, areas of land completely surrounded by water
- peninsulas, land surrounded by water on three sides

Because landforms play an important role in geography, many scientists study how landforms are made and how they affect human activity.

READING CHECK **Summarizing** What are some common landforms?

What You Will Learn...

Main Ideas

1. Earth's surface is covered by many different landforms.
2. Forces below Earth's surface build up our landforms.
3. Forces on the planet's surface shape Earth's landforms.
4. Landforms influence people's lives and culture.

The Big Idea

Processes below and on Earth's surface shape the planet's physical features.

Key Terms

landforms, *p. 35*
continents, *p. 36*
plate tectonics, *p. 36*
lava, *p. 37*
earthquakes, *p. 38*
weathering, *p. 39*
erosion, *p. 39*

TAKING NOTES As you read, use a diagram like the one below to take notes on Earth's landforms. In the circles, be sure to note how landforms are created, change, and affect people's lives.

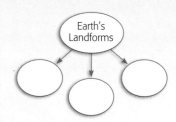

Forces below Earth's Surface

Geographers often study how landforms are made. One explanation for how landforms have been shaped involves forces below Earth's surface.

Earth's Plates

To understand how these forces work, we must examine Earth's **structure**. The planet is made up of three layers. A solid inner core is surrounded by a liquid layer, or mantle. The solid outer layer of Earth is called the crust. The planet's **continents**, or large landmasses, are part of Earth's crust.

Geographers use the theory of plate tectonics to explain how forces below Earth's surface have shaped our landforms. The theory of **plate tectonics** suggests that Earth's surface is divided into a dozen or so slow-moving plates, or pieces of Earth's crust. As you can see in the image below, some plates, like the Pacific plate, are quite large. Others, like the Nazca plate, are much smaller. These plates cover Earth's entire surface. Some plates are under the ocean. These are known as ocean plates. Other plates, known as continental plates, are under Earth's continents.

Why do these plates move? Energy deep inside the planet puts pressure on Earth's crust. As this pressure builds up, it forces the plates to shift. Earth's tectonic plates all move. However, they move in different directions and at different speeds.

The Movement of Continents

Earth's tectonic plates move slowly—up to several inches per year. The continents, which are part of Earth's plates, shift as the plates move. If we could look back some 200 million years, we would see that the continents have traveled great distances. This idea is known as continental drift.

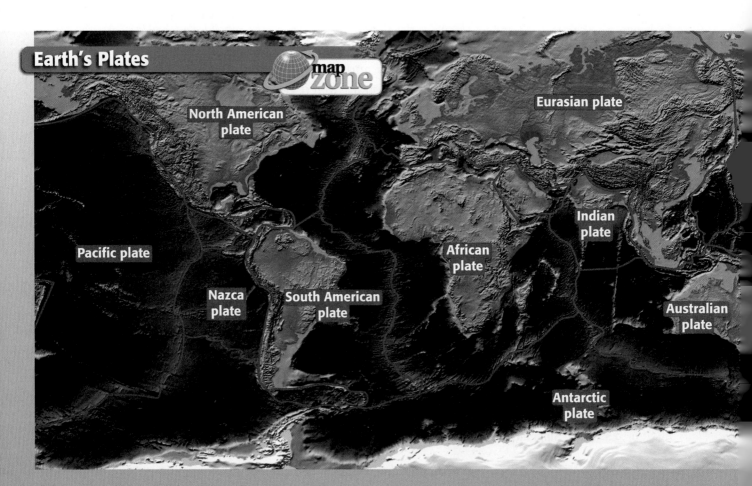

Earth's Plates

map zone

North American plate

Eurasian plate

Pacific plate

Nazca plate

South American plate

African plate

Indian plate

Australian plate

Antarctic plate

The theory of continental drift, first developed by Alfred Wegener, states that the continents were once united in a single supercontinent. According to this theory, Earth's plates shifted over millions of years. As a result, the continents slowly separated and moved to their present positions.

Earth's continents are still moving. Some plates move toward each other and collide. Other plates separate and move apart. Still others slide past one another. Over time, colliding, separating, and sliding plates have shaped Earth's landforms.

Plates Collide

As plates collide, the energy created from their collision produces distinct landforms. The collision of different types of plates creates different shapes on Earth's surface. Ocean trenches and mountain ranges are two examples of landforms produced by the collision of tectonic plates.

BIOGRAPHY

Alfred Wegener
(1880–1930)

German scientist Alfred Wegener's fascination with the similarities between the western coast of Africa and the eastern coast of South America led to his theory of continental drift. Wegener argued that the two continents had once been joined together. Years of plate movement broke the continents apart and moved them to their current locations. It was only after Wegener's death that his ideas became a central part of the theory of plate tectonics.

The theory of plate tectonics suggests that the plates that make up Earth's crust are moving, usually only a few inches per year. As Earth's plates collide, separate, and slide past each other, they create forces great enough to shape many of Earth's landforms.

ANALYZING VISUALS Looking at the map, what evidence indicates that plates have collided or separated?

When two ocean plates collide, one plate pushes under the other. This process creates ocean trenches. Ocean trenches are deep valleys in the ocean floor. Near Japan, for example, the Pacific plate is slowly moving under other plates. This collision has created several deep ocean trenches, including the world's deepest trench, the Mariana Trench.

Ocean plates and continental plates can also collide. When this occurs, the ocean plate drops beneath the continental plate. This action forces the land above to crumple and form a mountain range. The Andes in South America, for example, were formed when the South American and Nazca plates collided.

The collision of two continental plates also results in mountain-building. When continental plates collide, the land pushes up, sometimes to great heights. The world's highest mountain range, the Himalayas, formed when the Indian plate crashed into the Eurasian plate. In fact, the Himalayas are still growing as the two plates continue to crash into each other.

Plates Separate

A second type of plate movement causes plates to separate. As plates move apart, gaps between the plates allow magma, a liquid rock from the planet's interior, to rise to Earth's crust. **Lava**, or magma that reaches Earth's surface, emerges from the gap that has formed. As the lava cools, it builds a mid-ocean ridge, or underwater mountain. For example, the separation of the North American and Eurasian plates formed the largest underwater mountain, the Mid-Atlantic Ridge. If these mid-ocean ridges grow high enough, they can rise above the surface of the ocean, forming volcanic islands. Iceland, on the boundary of the Eurasian and North American plates, is an example of such an island.

FOCUS ON READING
The suffix –sion means the act of. What does the word collision mean?

Plates Slide

Tectonic plates also slide past each other. As plates pass by one another, they sometimes grind together. This grinding produces **earthquakes**—sudden, violent movements of Earth's crust. Earthquakes often take place along faults, or breaks in Earth's crust where movement occurs. In California, for example, the Pacific plate is sliding by the edge of the North American plate. This has created the San Andreas Fault zone, an area where earthquakes are quite common.

The San Andreas Fault zone is one of many areas that lie along the boundaries of the Pacific plate. The frequent movement of this plate produces many earthquakes and volcanic eruptions along its edges. In fact, the region around the Pacific plate, called the Ring of Fire, is home to most of the world's earthquakes and volcanoes.

READING CHECK **Finding Main Ideas** What forces below Earth's surface shape landforms?

Plate Movement

The movement of tectonic plates has produced many of Earth's landforms. Volcanoes, islands, and mountains often result from the separation or collision of Earth's plates.

ANALYZING VISUALS What type of landform is created by the collision of two continental plates?

The separation of plates can allow magma to rise up and create volcanic islands like Surtsey Island, near Iceland.

Plate A

Plate B

magma

The Himalayas in South Asia resulted from the collision of two massive continental plates.

Plate A

Plate B

Forces on Earth's Surface

For millions of years, the movement of Earth's tectonic plates has been building up landforms on Earth's surface. At the same time, other forces are working to change those very same landforms.

Imagine a small pile of dirt and rock on a table. If you poured water on the pile, it would move the dirt and rock from one place to another. Likewise, if you were to blow at the pile, the rock and dirt would also move. The same process happens in nature. Weather, water, and other forces change Earth's landforms by wearing them away or reshaping them.

Weathering

One force that wears away landforms is weathering. **Weathering** is the process by which rock is broken down into smaller pieces. Several factors cause rock to break down. In desert areas, daytime heating and nighttime cooling can cause rocks to crack. Water may get into cracks in rocks and freeze. The ice then expands with a force great enough to break the rock. Even the roots of trees can pry rocks apart.

Regardless of which weathering process is at work, rocks eventually break down. These small pieces of rock are known as sediment. Once weathering has taken place, wind, ice, and water often move sediment from one place to another.

Erosion

Another force that changes landforms is the process of erosion. **Erosion** is the movement of sediment from one location to another. Erosion can wear away or build up landforms. Wind, ice, and water all cause erosion.

Powerful winds often cause erosion. Winds lift sediment into the air and carry it across great distances. On beaches and in deserts, wind can deposit large amounts of sand to form dunes. Blowing sand can also wear down rock. The sand acts like sandpaper to polish and wear away at rocks. As you can see in the photo below, wind can have a dramatic effect on landforms.

Earth's glaciers also have the power to cause massive erosion. Glaciers, or large, slow-moving sheets of ice, build up when winter snows do not melt the following summer. Glaciers can be huge. Glaciers in Greenland and Antarctica, for example, are great sheets of ice up to two miles (3 km) thick. Some glaciers flow slowly downhill like rivers of ice. As they do so, they erode the land by carving large U-shaped valleys and sharp mountain peaks. As the ice flows downhill, it crushes rock into sediment and can move huge rocks long distances.

Wind Erosion

Landforms in Utah's Canyonlands National Park have been worn away by thousands of years of powerful winds.

39

Water is the most common cause of erosion. Waves in oceans and lakes can wear away the shore, creating jagged coastlines, like those on the coast of Oregon. Rivers also cause erosion. Over many years, the flowing water can cut through rock, forming canyons, or narrow areas with steep walls. Arizona's Horseshoe Bend and Grand Canyon are examples of canyons created in this way.

Flowing water shapes other landforms as well. When water deposits sediment in new locations, it creates new landforms. For example, rivers create floodplains when they flood their banks and deposit sediment along the banks. Sediment that is carried by a river all the way out to sea creates a delta. The sediment settles to the bottom, where the river meets the sea. The Nile and Mississippi rivers have created two of the world's largest river deltas.

READING CHECK **Comparing** How are weathering and erosion similar?

Water Erosion

For millions of years, the Colorado River has worn away the rock at Horseshoe Bend in Arizona.

Landforms Influence Life

Why do you live where you do? Perhaps your family moved to the desert to avoid harsh winter weather. Or possibly one of your ancestors settled near a river delta because its fertile soil was ideal for growing crops. Maybe your family wanted to live near the ocean to start a fishing business. As these examples show, landforms exert a strong influence on people's lives. Earth's landforms affect our settlements and our culture. At the same time, we affect the landforms around us.

Earth's landforms can influence where people settle. People sometimes settle near certain landforms and avoid others. For example, many settlements are built near fertile river valleys or deltas. The earliest urban civilization, for example, was built in the valley between the Tigris and Euphrates rivers. Other times, landforms discourage people from settling in a certain place. Tall, rugged mountains, like the Himalayas, and harsh desert climates, like the Sahara, do not usually attract large settlements.

Landforms affect our culture in ways that we may not have noticed. Landforms often influence what jobs are available in a region. For example, rich mineral deposits in the mountains of Colorado led to the development of a mining industry there. Landforms even affect language. On the island of New Guinea in Southeast Asia, rugged mountains have kept the people so isolated that more than 700 languages are spoken on the island today.

People sometimes change landforms to suit their needs. People may choose to modify landforms in order to improve their lives. For example, engineers built the Panama Canal to make travel from the Atlantic Ocean to the Pacific Ocean easier. In Southeast Asia, people who farm on steep hillsides cut terraces into the slope to

Living with Landforms

The people of Rio de Janeiro, Brazil, have learned to adapt to the mountains and bays that dominate their landscape.

ANALYZING VISUALS How have people in Rio de Janiero adapted to their landscape?

create more level space to grow their crops. People have even built huge dams along rivers to divert water for use in nearby towns or farms.

READING CHECK Analyzing What are some examples of humans adjusting to and changing landforms?

SUMMARY AND PREVIEW Landforms are created by actions deep within the planet's surface, and they are changed by forces on Earth's surface, like weathering and erosion. In the next chapter you will learn how other forces, like weather and climate, affect Earth's people.

Section 3 Assessment

go.hrw.com
Online Quiz
KEYWORD: SGA7 HP2

Reviewing Key Ideas, Terms, and Places

1. a. **Describe** What are some common **landforms**?
   b. **Analyze** Why do geographers study landforms?
2. a. **Identify** What is the theory of **plate tectonics**?
   b. **Compare and Contrast** How are the effects of colliding plates and separating plates similar and different?
   c. **Predict** How might Earth's surface change as tectonic plates continue to move?
3. a. **Recall** What is the process of **weathering**?
   b. **Elaborate** How does water affect sediment?
4. a. **Recall** How do landforms affect life on Earth?
   b. **Predict** How might people adapt to life in an area with steep mountains?

Critical Thinking

5. **Analyzing** Use your notes and the chart below to identify the different factors that alter Earth's landforms and the changes that they produce.

Factor	Change in Landform

FOCUS ON WRITING

6. **Writing about Earth's Land** Think of some vivid words you could use to describe Earth's landforms. As you think of them, add them to your notebook.

The Ring of Fire

Essential Elements

The World in Spatial Terms
Places and Regions
Physical Systems
Human Systems
Environment and Society
The Uses of Geography

Background Does "the Ring of Fire" sound like the title of a fantasy novel? It's actually the name of a region that circles the Pacific Ocean known for its fiery volcanoes and powerful earthquakes. The Ring of Fire stretches from the tip of South America all the way up to Alaska, and from Japan down to the islands east of Australia. Along this belt, the Pacific plate moves against several other tectonic plates. As a result, thousands of earthquakes occur there every year, and dozens of volcanoes erupt.

The Eruption of Mount Saint Helens One of the best-known volcanoes in the Ring of Fire is Mount Saint Helens in Washington State. Mount Saint Helens had been dormant, or quiet, since 1857. Then in March 1980, it began spitting out puffs of steam and ash. Officials warned people to leave the area. Scientists brought in equipment to measure the growing bulge in the mountainside. Everyone feared the volcano might erupt at any moment.

On May 18, after a sudden earthquake, Mount Saint Helens let loose a massive explosion of rock and lava. Heat from the blast melted snow on the mountain, which

Ring of Fire

map zone

Asia
North America
PACIFIC OCEAN
South America
Australia
Antarctica

Plate boundary line
Earthquakes
Active volcanoes

THE WORLD ALMANAC Facts about the World — Major Eruptions in the Ring of Fire

Volcano	Year
Tambora, Indonesia	1815
Krakatau, Indonesia	1883
Mount Saint Helens, United States	**1980**
Nevado del Ruiz, Colombia	1985
Mount Pinatubo, Philippines	1991

go.hrw.com KEYWORD: SGA7 CH2

Mount Saint Helens, 1980

The 1980 eruption of Mount Saint Helens blew ash and hot gases miles into the air. Today, scientists study the volcano to learn more about predicting eruptions.

mixed with ash to create deadly mudflows. As the mud quickly poured downhill, it flattened forests, swept away cars, and destroyed buildings. Clouds of ash covered the land, killing crops, clogging waterways, and blanketing towns as far as 200 miles (330 km) away. When the volcano finally quieted down, 57 people had died. Damage totaled nearly $1 billion. If it were not for the early evacuation of the area, the destruction could have been much worse.

What It Means By studying Mount Saint Helens, scientists learned a great deal about stratovolcanoes. These are tall, steep, cone-shaped volcanoes that have violent eruptions. Stratovolcanoes often form in areas where tectonic plates collide.

Because stratovolcanoes often produce deadly eruptions, scientists try to predict when they might erupt. The lessons learned from Mount Saint Helens helped scientists

warn people about another stratovolcano, Mount Pinatubo in the Philippines. That eruption in 1991 was the second-largest of the 1900s. It was far from the deadliest, however. Careful observation and timely warnings saved thousands of lives.

The Ring of Fire will always remain a threat. However, the better we understand its volcanoes, the better prepared we'll be when they erupt.

Geography for Life Activity

1. How did the eruption of Mount Saint Helens affect the surrounding area?

2. Why do scientists monitor volcanic activity?

3. **Investigating the Effects of Volcanoes** Some volcanic eruptions affect environmental conditions around the world. Research the eruption of either Mount Saint Helens or the Philippines' Mount Pinatubo to find out how its eruption affected the global environment.

Social Studies Skills

Chart and Graph   Critical Thinking   Geography   Study

Using a Physical Map

Learn

Physical maps show important physical features, like oceans and mountains, in a particular area. They also indicate an area's elevation, or the height of the land in relation to sea level.

When you use a physical map, there are important pieces of information you should always examine.

- Identify physical features. Natural features, such as mountains, rivers, and lakes, are labeled on physical maps. Read the labels carefully to identify what physical features are present.

- Read the legend. On physical maps, the legend indicates scale as well as elevation. The different colors in the elevation key indicate how far above or below sea level a place is.

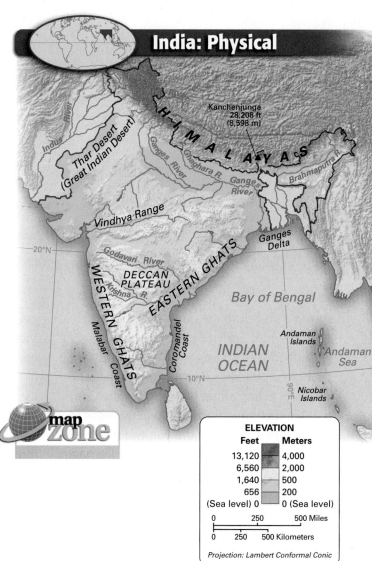

India: Physical

ELEVATION

Feet		Meters
13,120		4,000
6,560		2,000
1,640		500
656		200
(Sea level) 0		0 (Sea level)

0   250   500 Miles
0   250   500 Kilometers

Projection: Lambert Conformal Conic

Practice

Use the physical map of India at right to answer the questions below.

1. What landforms and bodies of water are indicated on the map?

2. What is the highest elevation in India? Where is it located?

Apply

Locate the physical map of Africa in the atlas in the back of the book. Use the map to answer the questions below.

1. Which region has the highest elevation?

2. What bodies of water surround Africa?

3. What large island is located off the east coast of Africa?

Chapter Review

Geography's Impact
video series
Review the video to answer the closing question:
What are some reasons for water shortages, and what can be done to solve this problem?

Visual Summary

Use the visual summary below to help you review the main ideas of the chapter.

QUICK FACTS

The amount of solar energy Earth receives changes based on Earth's movement and position.

Water is crucial to life on Earth. Our abundant water supply is stored in oceans, in lakes, and underground.

Earth's various landforms are shaped by complex processes both under and on the planet's surface.

Reviewing Vocabulary, Terms, and Places

For each statement below, write T if it is true and F if it is false. If the statement is false, write the correct term that would make the sentence a true statement.

1. **Weathering** is the movement of sediment from one location to another.

2. Because high **latitude** areas receive indirect rays from the sun, they have cooler temperatures.

3. Most of our **groundwater** is stored in Earth's streams, rivers, and lakes.

4. It takes 365¼ days for Earth to complete one **rotation** around the sun.

5. Streams are formed when **precipitation** collects in narrow channels.

6. **Earthquakes** cause erosion as they flow downhill, carving valleys and mountain peaks.

7. The planet's tilt affects the amount of **erosion** Earth receives from the sun.

Comprehension and Critical Thinking

SECTION 1 *(Pages 26–29)*

8. **a. Identify** What factors influence the amount of energy that different places on Earth receive from the sun?

   b. Analyze Why do the Northern and Southern hemispheres experience opposite seasons?

   c. Predict What might happen to the amount of solar energy we receive if Earth's axis were straight up and down?

SECTION 2 *(Pages 30–34)*

9. **a. Describe** What different sources of water are available on Earth?

   b. Draw Conclusions How does the water cycle keep Earth's water supply relatively constant?

   c. Elaborate What water problems affect people around the world? What solutions can you think of for one of those problems?

SECTION 3 *(Pages 35–41)*

10. a. Define What is a landform? What are some common types of landforms?

b. Analyze Why are Earth's landforms still changing?

c. Elaborate What physical features dominate the landscape in your community? How do they affect life there?

Using the Internet

go.hrw.com
KEYWORD: SGA7 CH2

11. Activity: Researching Earth's Seasons Earth's seasons not only affect temperatures, they also affect how much daylight is available during specific times of the year. Enter the activity keyword to research Earth's seasons and view animations to see how seasons change. Then use the interactive worksheet to answer some questions about what you learned.

FOCUS ON READING AND WRITING

Using Word Parts *Use what you learned about prefixes, suffixes, and word roots to answer the questions below.*

12. Examine the word *separation*. What is the suffix? What is the root? What does separation mean?

13. The prefix *in-* means not. What do the words *invisible* and *inactive* mean?

14. The suffix *-ment* means action or process. What does the word *movement* mean?

Writing a Haiku *Use your notes and the directions below to write a haiku.*

15. Look back through the notes you made about planet Earth. Choose one aspect of Earth to describe in a haiku. Haikus are short, three-line poems. Traditional haikus consist of only 17 syllables—five in the first line, seven in the second line, and five in the third line. You may choose to write a traditional haiku, or you may choose to write a haiku with a different number of syllables. Be sure to use descriptive words to paint a picture of planet Earth.

Social Studies Skills

Using a Physical Map *Examine the physical map of the United States in the back of this book. Use it to answer the questions below.*

16. What physical feature extends along the Gulf of Mexico?

17. What mountain range in the West lies above 6,560 feet?

18. Where does the elevation drop below sea level?

Map Activity

Physical Map *Use the map below to answer the questions that follow.*

19. Which letter indicates a river?

20. Which letter on the map indicates the highest elevation?

21. The lowest elevation on the map is indicated by which letter?

22. An island is indicated by which letter?

23. Which letter indicates a large body of water?

24. Which letter indicates an area between 1,640 feet and 6,560 feet above sea level?

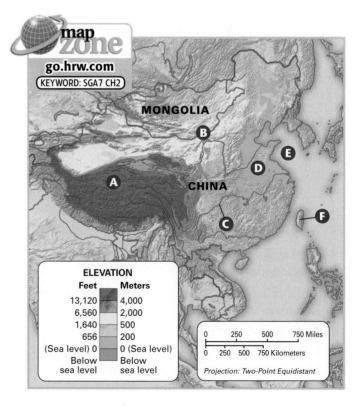

Standardized Test Practice

DIRECTIONS: Read questions 1 through 7 and write the letter of the best response. Then read question 8 and write your own well-constructed response.

1 **Which regions on Earth have seasons tied to the amount of rainfall?**

   A polar regions

   B the tropics

   C the Northern Hemisphere

   D high latitudes

2 **Most of Earth's water supply is made up of**

   A groundwater.

   B water vapor.

   C freshwater.

   D salt water.

3 **The theory of continental drift explains how**

   A Earth's continents have moved thousands of miles.

   B Earth's axis has moved to its current position.

   C mountains and valleys are formed.

   D sediment moves from one place to another.

4 **Which of the following is a cause of erosion?**

   A evaporation

   B ice

   C plate collisions

   D Earth's tilt

5 **Changes in solar energy that create day and night are a result of**

   A the movement of tectonic plates.

   B Earth's rotation.

   C the revolution of Earth around the sun.

   D Earth's tilt.

The Water Cycle

6 **In the illustration above, which letter *best* reflects the process of evaporation?**

   A W

   B X

   C Y

   D Z

7 **Which of the following is *most likely* a cause of water pollution?**

   A River water is used to produce electricity.

   B Heavy rainfall causes a river to overflow its banks.

   C Chemicals from a factory seep into the local water supply.

   D Groundwater is used faster than it can be replaced.

8 **Extended Response Question** Use the water cycle diagram above to explain how Earth's water cycle affects our water supply.

CHAPTER 3

Climate, Environment, and Resources

What You Will Learn...

In this chapter you will learn about weather and climate. Climate is the weather conditions over a long period of time. You will also learn about how living things and the environment are connected and about the importance of Earth's natural resources.

SECTION 1
Weather and Climate **50**

SECTION 2
World Climates **55**

SECTION 3
Natural Environments **62**

SECTION 4
Natural Resources **68**

FOCUS ON READING AND VIEWING

Understanding Cause and Effect A cause makes something happen. An effect is the result of a cause. Words such as *because, result, since,* and *therefore* can signal causes or effects. As you read, look for causes and effects to understand how things relate. **See the lesson, Understanding Cause and Effect, on page 108.**

Creating and Viewing a Weather Report You have likely seen a TV weather report, which tells the current weather conditions and predicts future conditions. After reading this chapter, prepare a weather report for a season and place of your choosing. Present your report to the class and then view your classmates' reports.

Climate Earth has many climates, such as the dry climate of the region shown here.

This photo shows a severe thunderstorm. These storms produce violent weather, such as heavy rainfall and strong winds, which affects people's lives.

How do you think this storm might have affected the people who lived in this area?

HOLT

Geography's Impact

video series
Watch the video to understand the impact of weather.

Environments Living things, such as this koala, depend on their surroundings.

Natural Resources Earth provides many valuable and useful natural resources, such as oil.

Weather and Climate

If YOU lived there...

You live in Buffalo, New York, at the eastern end of Lake Erie. One evening in January, you are watching the local TV news. The weather forecaster says, "A huge storm is brewing in the Midwest and moving east. As usual, winds from this storm will drop several feet of snow on Buffalo as they blow off Lake Erie."

Why will winds off the lake drop snow on Buffalo?

BUILDING BACKGROUND All life on Earth depends on the sun's energy and on the cycle of water from the land to the air and back again. In addition, sun and water work with other forces, such as wind, to create global patterns of weather and climate.

Understanding Weather and Climate

"Climate is what you expect; weather is what you get."
—Robert Heinlein, from *Time Enough for Love*

What is it like outside right now where you live? Is it hot, sunny, wet, cold? Is this what it is usually like outside for this time of year? The first two questions are about **weather**, the short-term changes in the air for a given place and time. The last question is about **climate**, a region's average weather conditions over a long period.

Weather is the temperature and precipitation from hour to hour or day to day. "Today is sunny, but tomorrow it might rain" is a statement about weather. Climate is the expected weather for a place based on data and experience. "Summer here is usually hot and muggy" is a statement about climate. The factors that shape weather and climate include the sun, location on Earth, wind, water, and mountains.

READING CHECK **Finding Main Ideas** How are weather and climate different from each other?

Sun and Location

Energy from the sun heats the planet. Different locations receive different amounts of sunlight, though. Thus, some locations are warmer than others. The differences are due to Earth's tilt, movement, and shape.

You have learned that Earth is tilted on its axis. The part of Earth tilted toward the sun receives more solar energy than the part tilted away from the sun. As the Earth revolves around the sun, the part of Earth that is tilted toward the sun changes during the year. This process creates the seasons. In general, temperatures in summer are warmer than in winter.

Earth's shape also affects the amount of sunlight different locations receive. Because Earth is a sphere, its surface is rounded. Therefore, solar rays are more direct and concentrated near the equator. Nearer the poles, the sun's rays are less direct and are spread over a larger region.

As a result, areas near the equator, called the lower latitudes, are mainly hot year-round. Areas near the poles, called the higher latitudes, are cold year-round. Areas about halfway between the equator and poles have more seasonal change. In general, the farther from the equator, or the higher the latitude, the colder the climate.

READING CHECK **Summarizing** How does Earth's tilt on its axis affect climate?

Wind and Water

Heat from the sun moves across Earth's surface. The reason is that air and water warmed by the sun are constantly on the move. You might have seen a gust of wind or a stream of water carrying dust or dirt. In a similar way, wind and water carry heat from place to place. As a result, they make different areas of Earth warmer or cooler.

Global Wind Systems

Prevailing winds blow in circular belts across Earth. These belts occur at about every 30° of latitude.

ANALYZING VISUALS Which direction do the prevailing winds blow across the United States?

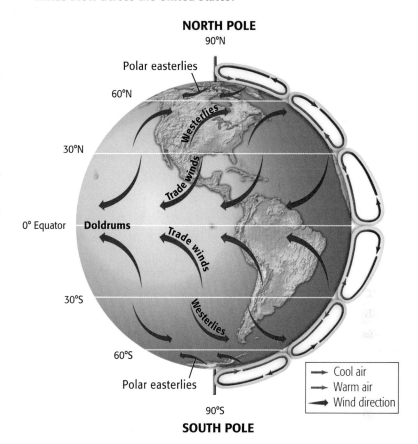

Global Winds

Wind, or the sideways movement of air, blows in great streams around the planet. **Prevailing winds** are winds that blow in the same direction over large areas of Earth. The diagram above shows the patterns of Earth's prevailing winds.

To understand Earth's wind patterns, you need to think about the weight of air. Although you cannot feel it, air has weight. This weight changes with the temperature. Cold air is heavier than warm air. For this reason, when air cools, it gets heavier and sinks. When air warms, it gets lighter and rises. As warm air rises, cooler air moves in to take its place, creating wind.

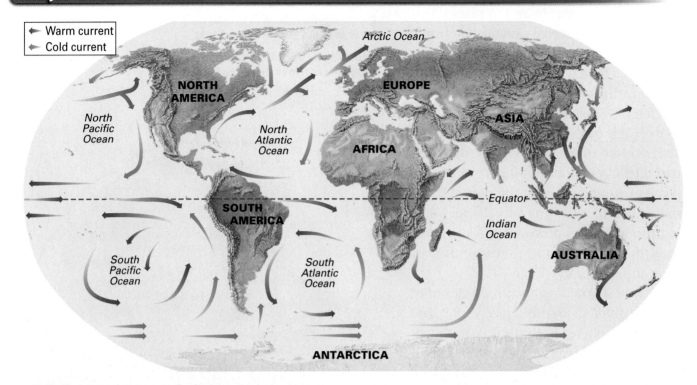

Warm current
Cold current

Arctic Ocean

NORTH AMERICA
EUROPE
ASIA
AFRICA
North Pacific Ocean
North Atlantic Ocean
SOUTH AMERICA
Equator
Indian Ocean
AUSTRALIA
South Pacific Ocean
South Atlantic Ocean
ANTARCTICA

Geography Skills

Movement Ocean currents carry warm water from the equator toward the poles and cold water from the poles toward the equator. The currents affect temperature.

1. **Use the Map** Does a warm or cold ocean current flow along the lower west coast of North America?
2. **Explain** How do ocean currents move heat between warmer and colder areas of Earth?

On a global scale, this rising, sinking, and flowing of air creates Earth's prevailing wind patterns. At the equator, hot air rises and flows toward the poles. At the poles, cold air sinks and flows toward the equator. Meanwhile, Earth is rotating. Earth's rotation causes prevailing winds to curve east or west rather than flowing directly north or south.

Depending on their source, prevailing winds make a region warmer or colder. In addition, the source of the winds can make a region drier or wetter. Winds that form from warm air or pass over lots of water often carry moisture. In contrast, winds that form from cold air or pass over lots of land often are dry.

FOCUS ON READING

What is the effect of Earth's rotation on prevailing winds?

Ocean Currents

Like wind, **ocean currents**—large streams of surface seawater—move heat around Earth. Winds drive these currents. The map above shows how Earth's ocean currents carry warm or cool water to different areas. The water's temperature affects air temperature near it. Warm currents raise temperatures; cold currents lower them.

The Gulf Stream is a warm current that flows north along the U.S. East Coast. It then flows east across the Atlantic to become the North Atlantic Drift. As the warm current flows along northwestern Europe, it heats the air. Westerlies blow the warmed air across Europe. This process makes Europe warmer than it otherwise would be.

Large Bodies of Water

Large bodies of water, such as an ocean or sea, also affect climate. Water heats and cools more slowly than land does. For this reason, large bodies of water make the temperature of the land nearby milder. Thus, coastal areas, such as the California coast, usually do not have as wide temperature ranges as inland areas.

As an example, the state of Michigan is largely surrounded by the Great Lakes. The lakes make temperatures in the state milder than other places as far north.

Wind, Water, and Storms

If you watch weather reports, you will hear about storms moving across the United States. Tracking storms is important to us because the United States has so many of them. As you will see, some areas of the world have more storms than others do.

Most storms occur when two air masses collide. An air mass is a large body of air. The place where two air masses of different temperatures or moisture content meet is a **front**. Air masses frequently collide in regions like the United States, where the westerlies meet the polar easterlies.

Fronts can produce rain or snow as well as severe weather such as thunderstorms and icy blizzards. Thunderstorms produce rain, lightning, and thunder. In the United States, they are most common in spring and summer. Blizzards produce strong winds and large amounts of snow and are most common during winter.

Thunderstorms and blizzards can also produce tornadoes, another type of severe storm. A tornado is a small, rapidly twisting funnel of air that touches the ground. Tornadoes usually affect a limited area and last only a few minutes. However, they can be highly destructive, uprooting trees and tossing large vehicles through the air. Tornadoes can be extremely deadly as well.

In 1925 a tornado that crossed Missouri, Illinois, and Indiana left 695 people dead. It is the deadliest U.S. tornado on record.

The largest and most destructive storms, however, are hurricanes. These large, rotating storms form over tropical waters in the Atlantic Ocean, usually from late summer to fall. Did you know that hurricanes and typhoons are the same? Typhoons are just hurricanes that form in the Pacific Ocean.

Extreme Weather

Severe weather is often dangerous and destructive. In the top photo, rescuers search for people during a flood in Yardley, Pennsylvania. Below, a tornado races across a wheat field.

53

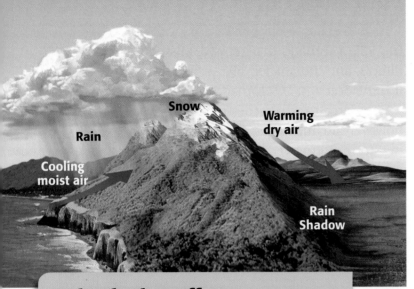

Rain

Snow

Warming dry air

Cooling moist air

Rain Shadow

Rain Shadow Effect

Most of the moisture in the ocean air falls on the mountainside facing the wind. LIttle moisture remains to fall on the other side, creating a rain shadow.

Mountains

Mountains can influence an area's climate by affecting both temperature and precipitation. Many high mountains are located in warm areas yet have snow at the top all year. How can this be? The reason is that temperature decreases with elevation—the height on Earth's surface above sea level.

Mountains also create wet and dry areas. Look at the diagram at left. A mountain forces air blowing against it to rise. As it rises, the air cools and precipitation falls as rain or snow. Thus, the side of the mountain facing the wind is often green and lush. However, little moisture remains for the other side. This effect creates a rain shadow, a dry area on the mountainside facing away from the direction of the wind.

READING CHECK Finding Main Ideas How does temperature change with elevation?

SUMMARY AND PREVIEW As you can see, the sun, location on Earth, wind, water, and mountains affect weather and climate. In the next section you will learn what the world's different climate regions are like.

Hurricanes produce drenching rain and strong winds that can reach speeds of 155 miles per hour (250 kph) or more. This is more than twice as fast as most people drive on highways. In addition, hurricanes form tall walls of water called storm surges. When a storm surge smashes into land, it can wipe out an entire coastal area.

READING CHECK Analyzing Why do coastal areas have milder climates than inland areas?

Section 1 Assessment

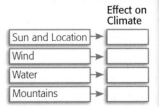

go.hrw.com
Online Quiz
KEYWORD: SGA7 HP3

Reviewing Key Ideas, Terms, and Places

1. a. **Recall** What shapes **weather** and **climate**?
   b. **Contrast** How do weather and climate differ?
2. a. **Identify** What parts of Earth receive the most heat from the sun?
   b. **Explain** Why do the poles receive less solar energy than the equator does?
3. a. **Describe** What creates wind?
   b. **Summarize** How do **ocean currents** and large bodies of water affect climate?
4. a. **Define** What is a rain shadow?
   b. **Explain** Why might a mountaintop and a nearby valley have widely different temperatures?

Critical Thinking

5. **Identifying Cause and Effect** Draw a chart like this one. Use your notes to explain how each factor affects climate.

	Effect on Climate
Sun and Location →	
Wind →	
Water →	
Mountains →	

FOCUS ON VIEWING

6. **Writing about Weather and Climate** Jot down information to include in your weather report. For example, you might want to include a term such as *fronts* or describe certain types of storms such as hurricanes or tornadoes.

World Climates

If YOU lived there...

You live in Colorado and are on your first serious hike in the Rocky Mountains. Since it is July, it is hot in the campground in the valley. But your guide insists that you bring a heavy fleece jacket. By noon, you have climbed to 11,000 feet. You are surprised to see patches of snow in shady spots. Suddenly, you are very happy that you brought your jacket!

Why does it get colder as you climb higher?

BUILDING BACKGROUND While weather is the day-to-day changes in a certain area, climate is the average weather conditions over a long period. Earth's different climates depend partly on the amount of sunlight a region receives. Differences in climate also depend on factors such as wind, water, and elevation.

Major Climate Zones

In January, how will you dress for the weekend? In some places, you might get dressed to go skiing. In other places, you might head out in a swimsuit to go to the beach. What the seasons are like where you live depends on climate.

Earth is a patchwork of climates. Geographers identify these climates by looking at temperature, precipitation, and native plant life. Using these items, we can divide Earth into five general climate zones—tropical, temperate, polar, dry, and highland.

The first three climate zones relate to latitude. Tropical climates occur near the equator, in the low latitudes. Temperate climates occur about halfway between the equator and the poles, in the middle latitudes. Polar climates occur near the poles, in the high latitudes. The last two climate zones occur at many different latitudes. In addition, geographers divide some climate zones into more specific climate regions. The chart and map on the next two pages describe the world's climate regions.

READING CHECK **Drawing Inferences** Why do you think geographers consider native plant life when categorizing climates?

What You Will Learn...

Main Ideas

1. Geographers use temperature, precipitation, and plant life to identify climate zones.
2. Tropical climates are wet and warm, while dry climates receive little or no rain.
3. Temperate climates have the most seasonal change.
4. Polar climates are cold and dry, while highland climates change with elevation.

The Big Idea

Earth's five major climate zones are identified by temperature, precipitation, and plant life.

Key Terms

monsoons, *p. 58*
savannas, *p. 58*
steppes, *p. 59*
permafrost, *p. 61*

TAKING NOTES As you read, use a chart like the one here to help you note the characteristics of Earth's major climate zones.

Climate Zone	Characteristics

World Climate Regions

To explore the world's climate regions, start with the chart below. After reading about each climate region, locate the places on the map that have that climate. As you locate climates, look for patterns. For example, places near the equator tend to have warmer climates than places near the poles. See if you can identify some other climate patterns.

Tropical climate

Climate		Where is it?	What is it like?	Plants
Tropical	HUMID TROPICAL	On and near the equator	Warm with high amounts of rain year-round; in a few places, monsoons create extreme wet seasons	Tropical rain forest
	TROPICAL SAVANNA	Higher latitudes in the tropics	Warm all year; distinct rainy and dry seasons; at least 20 inches (50 cm) of rain during the summer	Tall grasses and scattered trees
Dry	DESERT	Mainly center on 30° latitude; also in middle of continents, on west coasts, or in rain shadows	Sunny and dry; less than 10 inches (25 cm) of rain a year; hot in the tropics; cooler with wide daytime temperature ranges in middle latitudes	A few hardy plants, such as cacti
	STEPPE	Mainly bordering deserts and interiors of large continents	About 10–20 inches (25–50 cm) of precipitation a year; hot summers and cooler winters with wide temperature ranges during the day	Shorter grasses; some trees and shrubs by water
Temperate	MEDITERRANEAN	West coasts in middle latitudes	Dry, sunny, warm summers; mild, wetter winters; rain averages 15–20 inches (30–50 cm) a year	Scrub woodland and grassland
	HUMID SUBTROPICAL	East coasts in middle latitudes	Humid with hot summers and mild winters; rain year-round; in paths of hurricanes and typhoons	Mixed forest
	MARINE WEST COAST	West coasts in the upper-middle latitudes	Cloudy, mild summers and cool, rainy winters; strong ocean influence	Evergreen forests
	HUMID CONTINENTAL	East coasts and interiors of upper-middle latitudes	Four distinct seasons; long, cold winters and short, warm summers; average precipitation varies	Mixed forest

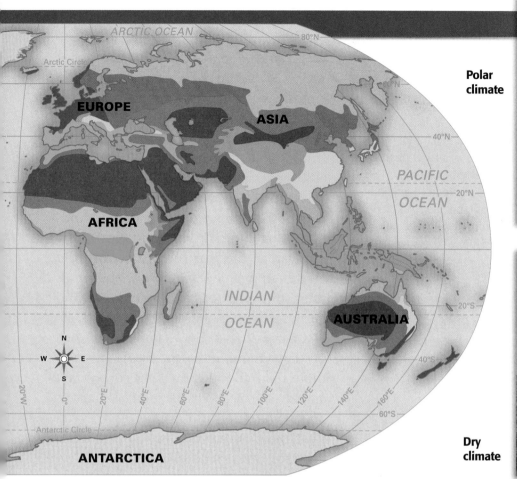

Polar climate

Dry climate

Climate		Where is it?	What is it like?	Plants
Polar	SUBARCTIC	Higher latitudes of the interior and east coasts of continents	Extremes of temperature; long, cold winters and short, warm summers; little precipitation	Northern evergreen forests
	TUNDRA	Coasts in high latitudes	Cold all year; very long, cold winters and very short, cool summers; little precipitation; permafrost	Moss, lichens, low shrubs
	ICE CAP	Polar regions	Freezing cold; snow and ice; little precipitation	No vegetation
Highland	HIGHLAND	High mountain regions	Wide range of temperatures and precipitation amounts, depending on elevation and location	Ranges from forest to tundra

map zone
Geography Skills

Regions Note how Earth's climate regions relate to different locations.
1. **Locate** Which climates are found mainly in the Northern Hemisphere?
2. **Identify** What climate does most of northern Africa have?

3. **Make Generalizations** Where are many of the world's driest climates found on Earth?
4. **Interpreting Charts** Examine the chart. Which two climates have the least amount of vegetation?

go.hrw.com (KEYWORD: SGA7 CH3)

The Tuareg of the Sahara.

In the Sahara, the world's largest desert, temperatures can top 130°F (54°C). Yet the Tuareg (TWAH-reg) of North and West Africa call the Sahara home—and prefer it. The Tuareg have raised camels and other animals in the Sahara for more than 1,000 years. The animals graze on sparse desert plants. When the plants are gone, the Tuareg move on.

In camp, Tuareg families live in tents made from animal skins. Some wealthier Tuareg live in adobe homes. The men traditionally wear blue veils wrapped around their face and head. The veils help protect against windblown desert dust.

Summarizing How have the Tuareg adapted to life in a desert?

Tropical and Dry Climates

Are you the type of person who likes to go to extremes? Then tropical and dry climates might be for you. These climates include the wettest, driest, and hottest places on Earth.

Tropical Climates

Our tour of Earth's climates starts at the equator, in the heart of the tropics. This region extends from the Tropic of Cancer to the Tropic of Capricorn. Look back at the map to locate this region.

Humid Tropical Climate At the equator, the hot, damp air hangs like a thick, wet blanket. Sweat quickly coats your body.

Welcome to the humid tropical climate. This climate is warm, muggy, and rainy year-round. Temperatures average about 80°F (26°C). Showers or storms occur almost daily, and rainfall ranges from 70 to more than 450 inches (180 to 1,140 cm) a year. In comparison, only a few parts of the United States average more than 70 inches (180 cm) of rain a year.

Some places with a humid tropical climate have **monsoons**, seasonal winds that bring either dry or moist air. During one part of the year, a moist ocean wind creates an extreme wet season. The winds then shift direction, and a dry land wind creates a dry season. Monsoons affect several parts of Asia. For example, the town of Mawsynram, India, receives on average more than 450 inches (1,140 cm) of rain a year—all in about six months! That is about 37 feet (11 m) of rain. As you can imagine, flooding during wet seasons is common and can be severe.

The humid tropical climate's warm temperatures and heavy rainfall support tropical rain forests. These lush forests contain more types of plants and animals than anywhere else on Earth. The world's largest rain forest is in the Amazon River basin in South America. There you can find more than 50,000 species, including giant lily pads, poisonous tree frogs, and toucans.

Tropical Savanna Climate Moving north and south away from the equator, we find the tropical savanna climate. This climate has a long, hot, dry season followed by short periods of rain. Rainfall is much lower than at the equator but still high. Temperatures are hot in the summer, often as high as 90°F (32°C). Winters are cooler but rarely get cold.

This climate does not receive enough rainfall to support dense forests. Instead, it supports **savannas**—areas of tall grasses and scattered trees and shrubs.

Dry Climates

Leaving Earth's wettest places, we head to its driest. These climates are found in a number of locations on the planet.

Desert Climate Picture the sun baking down on a barren wasteland. This is the desert, Earth's hottest and driest climate. Deserts receive less than 10 inches (25 cm) of rain a year. Dry air and clear skies produce high daytime temperatures and rapid cooling at night. In some deserts, highs can top 130°F (54°C)! Under such conditions, only very hardy plants and animals can live. Many plants grow far apart so as not to compete for water. Others, such as cacti, store water in fleshy stems and leaves.

Steppe Climate Semidry grasslands or prairies—called **steppes** (STEPS)—often border deserts. Steppes receive slightly more rain than deserts do. Short grasses are the most common plants, but shrubs and trees grow along streams and rivers.

READING CHECK **Contrasting** What are some ways in which tropical and dry climates differ?

Temperate Climates

If you enjoy hot, sunny days as much as chilly, rainy ones, then temperate climates are for you. *Temperate* means "moderate" or "mild." These mild climates tend to have four seasons, with warm or hot summers and cool or cold winters.

Temperate climates occur in the middle latitudes, the regions halfway between the equator and the poles. Air masses from the tropics and the poles often meet in these regions, which creates a number of different temperate climates. You very likely live in one, because most Americans do.

Mediterranean Climate Named for the region of the Mediterranean Sea, this sunny, pleasant climate is found in many popular vacation areas. In a Mediterranean climate, summers are hot, dry, and sunny. Winters are mild and somewhat wet. Plant life includes shrubs and short trees with scattered larger trees. The Mediterranean climate occurs mainly in coastal areas. In the United States, much of California has this climate.

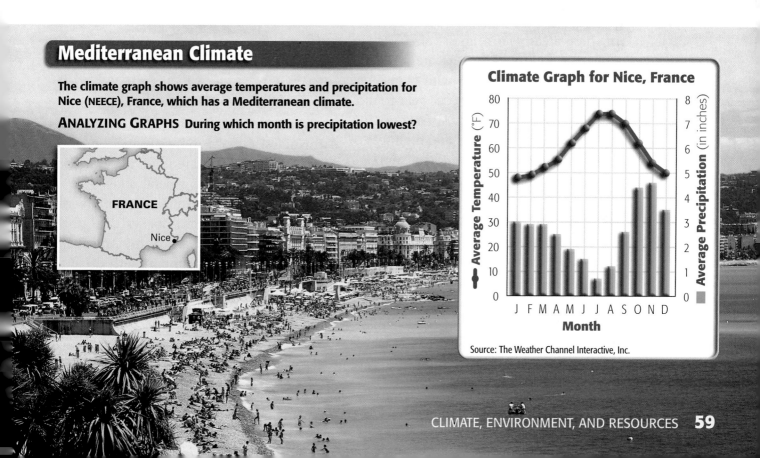

Mediterranean Climate

The climate graph shows average temperatures and precipitation for Nice (NEECE), France, which has a Mediterranean climate.

ANALYZING GRAPHS During which month is precipitation lowest?

FRANCE
Nice

Climate Graph for Nice, France

Source: The Weather Channel Interactive, Inc.

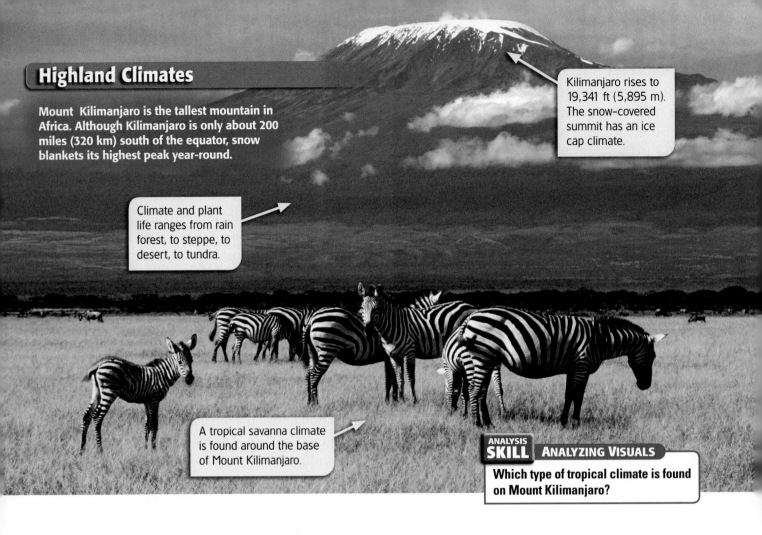

Highland Climates

Mount Kilimanjaro is the tallest mountain in Africa. Although Kilimanjaro is only about 200 miles (320 km) south of the equator, snow blankets its highest peak year-round.

Kilimanjaro rises to 19,341 ft (5,895 m). The snow-covered summit has an ice cap climate.

Climate and plant life ranges from rain forest, to steppe, to desert, to tundra.

A tropical savanna climate is found around the base of Mount Kilimanjaro.

ANALYSIS SKILL **ANALYZING VISUALS**

Which type of tropical climate is found on Mount Kilimanjaro?

Humid Subtropical Climate The southeastern United States is an example of the humid subtropical climate. This climate occurs along east coasts near the tropics. In these areas, warm, moist air blows in from the ocean. Summers are hot and muggy. Winters are mild, with occasional frost and snow. Storms occur year-round. In addition, hurricanes can strike, bringing violent winds, heavy rain, and high seas.

A humid subtropical climate supports mixed forests. These forests include both deciduous trees, which lose their leaves each fall, and coniferous trees, which are green year-round. Coniferous trees are also known as evergreens.

Marine West Coast Climate Parts of North America's Pacific coast and of western Europe have a marine west coast climate. This climate occurs on west coasts where winds carry moisture in from the seas.

The moist air keeps temperatures mild year-round. Winters are foggy, cloudy, and rainy, while summers can be warm and sunny. Dense evergreen forests thrive in this climate.

Humid Continental Climate Closer to the poles, in the upper–middle latitudes, many inland and east coast areas have a humid continental climate. This climate has four **distinct** seasons. Summers are short and hot. Spring and fall are mild, and winters are long, cold, and in general, snowy.

This climate's rainfall supports vast grasslands and forests. Grasses can grow very tall, such as in parts of the American Great Plains. Forests contain both deciduous and coniferous trees, with coniferous forests occurring in the colder areas.

READING CHECK **Categorizing** Which of the temperate climates is too dry to support forests?

ACADEMIC VOCABULARY

distinct clearly different and separate

Polar and Highland Climates

Get ready to feel the chill as we end our tour in the polar and highland climates. The three polar climates are found in the high latitudes near the poles. The varied highland climate is found on mountains.

Subarctic Climate The subarctic climate and the tundra climate described below occur mainly in the Northern Hemisphere south of the Arctic Ocean. In the subarctic climate, winters are long and bitterly cold. Summers are short and cool. Temperatures stay below freezing for about half the year. The climate's moderate rainfall supports vast evergreen forests called taiga (TY-guh).

Tundra Climate The tundra climate occurs in coastal areas along the Arctic Ocean. As in the subarctic climate, winters are long and bitterly cold. Temperatures rise above freezing only during the short summer. Rainfall is light, and only plants such as mosses, lichens, and small shrubs grow.

In parts of the tundra, soil layers stay frozen all year. Permanently frozen layers of soil are called **permafrost**. Frozen earth absorbs water poorly, which creates ponds and marshes in summer. This moisture causes plants to burst forth in bloom.

Ice Cap Climate The harshest places on Earth may be the North and South poles. These regions have an ice cap climate. Temperatures are bone-numbingly cold, and lows of more than –120°F (–84°C) have been recorded. Snow and ice remain year-round, but precipitation is light. Not surprisingly, no vegetation grows. However, mammals such as penguins and polar bears thrive.

Highland Climates The highland climate includes polar climates plus others. In fact, this mountain climate is actually several climates in one. As you go up a mountain, the climate changes. Temperatures drop, and plant life grows sparser. Going up a mountain can be like going from the tropics to the poles. On very tall mountains, ice coats the summit year-round.

FOCUS ON READING
What is the effect of elevation on climate?

READING CHECK **Comparing** How are polar and highland climates similar?

SUMMARY AND PREVIEW As you can see, Earth has many climates, which we identify based on temperature, precipitation, and native plant life. In the next section you will read about how nature and all living things are connected.

Section 2 Assessment

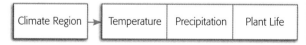
go.hrw.com
Online Quiz
KEYWORD: SGA7 HP3

Reviewing Key Ideas, Terms, and Places

1. **a. Recall** Which three major climate zones occur at certain latitudes?
   b. Summarize How do geographers categorize Earth's different climates?
2. **a. Define** What are **monsoons**?
   b. Make Inferences In which type of dry climate do you think the fewest people live, and why?
3. **a. Identify** What are the four temperate climates?
   b. Draw Conclusions Why are places with a Mediterranean climate popular vacation spots?
4. **a. Describe** What are some effects of **permafrost**?
   b. Explain How are highland climates unique?

Critical Thinking

5. **Categorizing** Create a chart like the one below for each climate region. Then use your notes to describe each climate region's average temperatures, precipitation, and native plant life.

Climate Region	→	Temperature	Precipitation	Plant Life

FOCUS ON VIEWING

6. **Discussing World Climates** Add information about the climate of the place you have selected, such as average temperature and precipitation.

Natural Environments

What You Will Learn...

Main Ideas

1. The environment and life are interconnected and exist in a fragile balance.
2. Soils play an important role in the environment.

The Big Idea

Plants, animals, and the environment, including soil, interact and affect one another.

Key Terms

environment, *p. 62*
ecosystem, *p. 63*
habitat, *p. 64*
extinct, *p. 64*
humus, *p. 65*
desertification, *p. 65*

 TAKING NOTES As you read, use a chart like the one below to help you take notes on the main topics in this section.

Limits and Connections in Nature	
Changes to Environments	
Soil and the Environment	

If YOU lived there...

When your family moved to the city, you were sure you would miss the woods and pond near your old house. Then one of your new friends at school told you there's a large park only a few blocks away. You wondered how interesting a city park could be. But you were surprised at the many plants and animals that live there.

What environments might you see in the park?

BUILDING BACKGROUND No matter where you live, you are part of a natural environment. From a desert to a rain forest to a city park, every environment is home to a unique community of plant and animal life. These plants and animals live in balance with nature.

The Environment and Life

If you saw a wild polar bear outside your school, you would likely be shocked. In most parts of the United States, polar bears live only in zoos. This is because plants and animals must live where they are suited to the **environment**, or surroundings. Polar bears are suited to very cold places with lots of ice, water, and fish. As you will see, living things and their environments are connected and affect each other in many ways.

Limits on Life

The environment limits life. As our tour of the world's climates showed, factors such as temperature, rainfall, and soil conditions limit where plants and animals can live. Palm trees cannot survive at the frigid North Pole. Ferns will quickly wilt and die in deserts, but they thrive in tropical rain forests.

At the same time, all plants and animals are adapted to specific environments. For example, kangaroo rats are adapted to dry desert environments. These small rodents can get all the water they need from food, so they seldom have to drink water.

Connections in Nature

The interconnections between living things and the environment form ecosystems. An **ecosystem** is a group of plants and animals that depend on each other for survival and the environment in which they live. Ecosystems can be any size and can occur wherever air, water, and soil support life. A garden pond, a city park, a prairie, and a rain forest are all examples of ecosystems.

The diagram below shows a forest ecosystem. Each part of this ecosystem fills a certain role. The sun provides energy to the plants, which use the energy to make their own food. The plants then serve as food, either directly or indirectly, for all other life in the forest. When the plants and animals die, their remains break down and provide nutrients for the soil and new plant growth. Thus, the cycle continues.

Close-up

A Forest Ecosystem

A forest is one type of ecosystem. The plants and animals in the forest depend on one another and the forest environment for survival.

1 Sunlight is the source of energy for most living things.

2 Plants use the energy in sunlight to make food. They serve as the basis for other life in the ecosystem.

3 Animals such as rabbits eat plants and gain some of their energy.

4 Predators, such as wolves and hawks, eat rabbits and other prey for energy.

5 Larger predators, such as mountain lions, compete for the prey that is available.

ANALYSIS SKILL **ANALYZING VISUALS**

What might happen in the forest ecosystem above if the number of rabbits fell significantly?

Changes to Environments

The interconnected parts of an ecosystem exist in a fragile balance. For this reason, a small change to one part can affect the whole system. A lack of rain in the forest ecosystem could kill off many of the plants that feed the rabbits. If the rabbits die, there will be less food for the wolves and mountain lions. Then they too may die.

Many actions can affect ecosystems. For example, people need places to live and food to eat, so they clear land for homes and farms. Clearing land has **consequences**, however. It can cause the soil to erode. In addition, the plants and animals that live in the area might be left without food and shelter.

Actions such as clearing land and polluting can destroy habitats. A **habitat** is the place where a plant or animal lives. The most diverse habitats on Earth are tropical rain forests. People are clearing Earth's rain forests for farmland, lumber, and other reasons, though. As a result, these diverse habitats are being lost.

FOCUS ON READING

What are some causes of habitat destruction?

Extreme changes in ecosystems can cause species to die out, or become **extinct**. As an example, flightless birds called dodos once lived on Mauritius (maw-RI-shuhs), an island in the Indian Ocean. When people began settling on the island, their actions harmed the dodos' habitat. First seen in 1507, dodos were extinct by 1681.

Recognizing these problems, many countries are working to balance people's needs with the needs of the environment. The United States, for example, has passed many laws to limit pollution, manage forests, and protect valuable ecosystems.

READING CHECK Drawing Inferences How might one change affect an entire ecosystem?

Soil and the Environment

As you know, plants are the basis for all food that animals eat. Soils help determine what plants will grow and how well. Because soils support plant life, they play an important role in the environment.

CONNECTING TO Science

Soil Factory

The next time you see a fallen tree in the forest, do not think of it as a dead log. Think of it as a soil factory. A fallen tree is buzzing with the activity of countless insects, bacteria, and other organisms. These organisms invade the fallen log and start to break the wood down.

As the tree decays and crumbles, it turns into humus. Humus is a rich blend of organic material. The humus mixes with the soil and adds valuable nutrients to it. These nutrients then enrich the soil, making it possible for new trees and plants to grow. Fallen trees provide as much as one-third of the organic material in forest soil.

Summarizing What causes a fallen tree to change into soil?

Fertile soils are rich in minerals and **humus** (HYOO-muhs), decayed plant or animal matter. These soils can support abundant plant life. Like air and water, fertile soil is essential for life. Without it, we could not grow much of the food we eat.

Soils can lose fertility in several ways. Erosion from wind or water can sweep topsoil away. Planting the same crops over and over can also rob soil of its fertility. When soil becomes worn out, it cannot support as many plants. In fragile dry environments this can lead to the spread of desertlike conditions, or **desertification**. The spread of desertlike conditions is a serious problem in many parts of the world.

READING CHECK **Analyzing** What do fertile soils contain, and why are these soils important?

SUMMARY AND PREVIEW Living things and the environment are connected, but changes can easily upset the balance in an ecosystem. Because they support plant life, soils are important parts of ecosystems. In the next section you will learn about Earth's many resources.

Soil Layers

The three layers of soil are the topsoil, subsoil, and broken rock. The thickness of each layer depends on the conditions in a specific location.

ANALYZING VISUALS In which layer of soil are most plant roots and insects found?

Topsoil

Subsoil

Broken Rock

Solid Rock

Section 3 Assessment

go.hrw.com
Online Quiz
KEYWORD: SGA7 HP3

Reviewing Key Ideas, Terms, and Places

1. **a. Define** What is an **ecosystem**, and what are two examples of ecosystems?
   b. Summarize How do nature and people change ecosystems?
   c. Elaborate Why can plants and animals not live everywhere?
2. **a. Recall** What is **humus**, and why is it important to soil?
   b. Identify Cause and Effect What actions can cause **desertification**, and what might be some possible effects?
   c. Elaborate Why it is important for geographers and scientists to study soils?

Critical Thinking

3. **Identifying Cause and Effect** Review your notes. Then use a chart like this one to identify some of the causes and effects of changes to ecosystems.

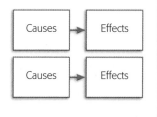

FOCUS ON VIEWING

4. **Writing about Natural Environments** Jot down ideas about how different types of weather might affect the environment of the place you chose. For example, how might lack of rain affect the area?

Earth's Changing Environments

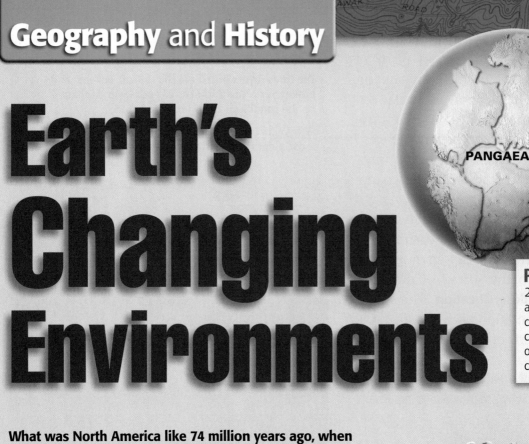

PANGAEA

Pangaea About 250 million years ago, all of Earth's continents were connected, forming one giant landmass called Pangaea.

What was North America like 74 million years ago, when dinosaurs roamed Earth? You might be surprised to learn that it was a very different place. Earth's environments are always changing. The map at right shows North America in the age of dinosaurs. Back then, the climate was warm and humid, and large inland seas covered much of the land. The region's plants and animals were completely different. Slowly, however, things changed. Some major event, possibly an asteroid impact, wiped out the dinosaurs. Over time, North America's environments changed into the ones that exist today.

What Survived Dinosaurs, such as the plant-eating ceratopsian at left, are long gone. But insects, such as cockroaches and dragon-flies, are still around.

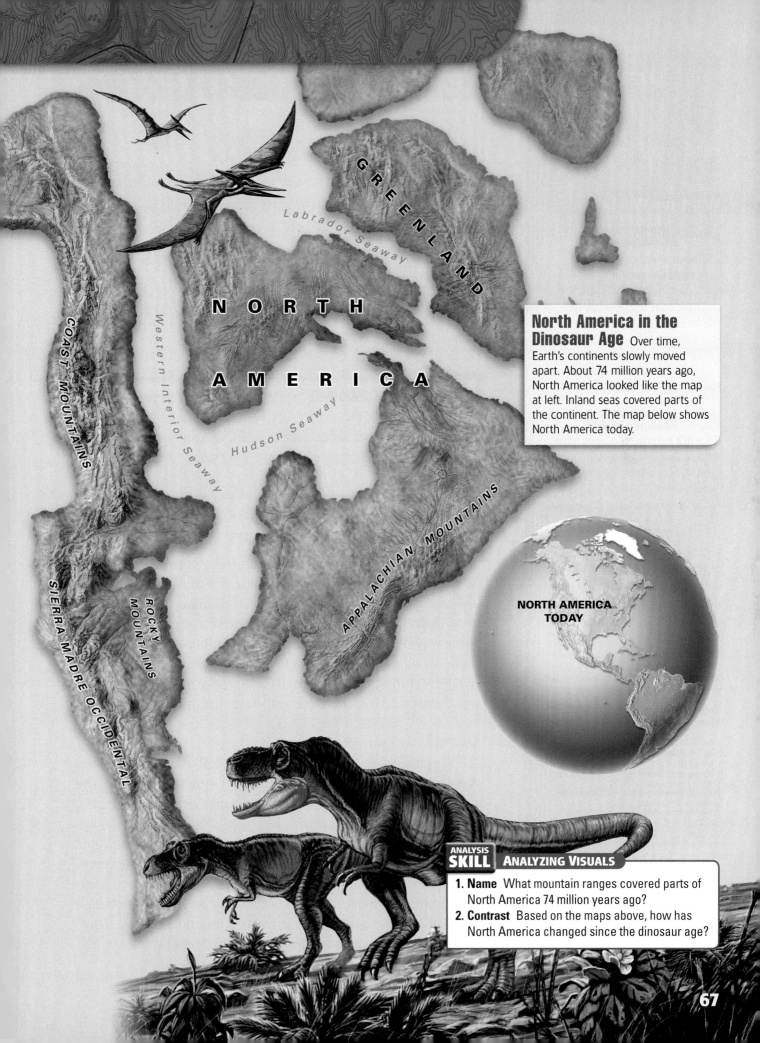

G R E E N L A N D

Labrador Seaway

N O R T H

A M E R I C A

COAST MOUNTAINS

Western Interior Seaway

Hudson Seaway

APPALACHIAN MOUNTAINS

SIERRA MADRE OCCIDENTAL

ROCKY MOUNTAINS

North America in the Dinosaur Age Over time, Earth's continents slowly moved apart. About 74 million years ago, North America looked like the map at left. Inland seas covered parts of the continent. The map below shows North America today.

NORTH AMERICA TODAY

ANALYSIS SKILL **ANALYZING VISUALS**

1. **Name** What mountain ranges covered parts of North America 74 million years ago?
2. **Contrast** Based on the maps above, how has North America changed since the dinosaur age?

Natural Resources

What You Will Learn...

Main Ideas

1. Earth provides valuable resources for our use.
2. Energy resources provide fuel, heat, and electricity.
3. Mineral resources include metals, rocks, and salt.
4. Resources shape people's lives and countries' wealth.

The Big Idea

Earth's natural resources have many valuable uses, and their availability affects people in many ways.

Key Terms

natural resource, *p. 68*
renewable resources, *p. 69*
nonrenewable resources, *p. 69*
deforestation, *p. 69*
reforestation, *p. 69*
fossil fuels, *p. 69*
hydroelectric power, *p. 70*

TAKING NOTES As you read, use a chart like this one to take notes on Earth's resources.

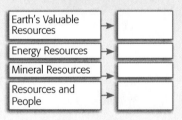

Earth's Valuable Resources	
Energy Resources	
Mineral Resources	
Resources and People	

If YOU lived there...

You live in Southern California, where the climate is warm and dry. Every week, you water the grass around your house to keep it green. Now the city has declared a "drought emergency" because of a lack of rain. City officials have put limits on watering lawns and on other uses of water.

How can you help conserve scarce water?

BUILDING BACKGROUND In addition to plant and animal life, other resources in the environment greatly influence people. In fact, certain vital resources, such as water, soils, and minerals, may determine whether people choose to live in a place or how wealthy people are.

Earth's Valuable Resources

Think about the materials in nature that you use. You have learned about the many ways we use sun, water, and land. They are just a start, though. Look at the human-made products around you. They all required the use of natural materials in some way. We use trees to make paper for books. We use petroleum, or oil, to make plastics for cell phones. We use metals to make machines, which we then use to make many items. Without these materials, our lives would change drastically.

Using Natural Resources

Trees, oil, and metals are all examples of natural resources. A **natural resource** is any material in nature that people use and value. Earth's most important natural resources include air, water, soils, forests, and minerals.

Understanding how and why people use natural resources is an important part of geography. We use some natural resources just as they are, such as wind. Usually, though, we change natural resources to make something new. For example, we change metals to make products such as bicycles and watches. Thus, most natural resources are raw materials for other products.

Reforestation

Members of the Green Belt Movement plant trees in Kenya. Although trees are a renewable resource, some forests are being cut down faster than new trees can replace them. Reforestation helps protect Earth's valuable forestlands.

ANALYZING VISUALS How does reforestation help the environment?

Types of Natural Resources

We group natural resources into two types, those we can replace and those we cannot. **Renewable resources** are resources Earth replaces naturally. For example, when we cut down a tree, another tree can grow in its place. Renewable resources include water, soil, trees, plants, and animals. These resources can last forever if used wisely.

Other natural resources will run out one day. These **nonrenewable resources** are resources that cannot be replaced. For example, coal formed over millions of years. Once we use the coal up, it is gone.

Managing Natural Resources

People need to manage natural resources to protect them for the future. Consider how your life might change if we ran out of forests, for example. Although forests are renewable, we can cut down trees far faster than they can grow. The result is the clearing of trees, or **deforestation**.

By managing resources, however, we can repair and prevent resource loss. For example, some groups are engaged in **reforestation**, planting trees to replace lost forestland.

READING CHECK Contrasting How do renewable and nonrenewable resources differ?

BIOGRAPHY

Wangari Maathai
(1940–)

Can planting a tree improve people's lives? Wangari Maathai thinks so. Born in Kenya in East Africa, Maathai wanted to help people in her country, many of whom were poor. She asked herself what Kenyans could do to improve their lives. "Planting a tree was the best idea that I had," she says. In 1977 Maathai founded the Green Belt Movement to plant trees and protect forestland. The group has now planted more than 30 million trees across Kenya! These trees provide wood and prevent soil erosion. In 2004 Maathai was awarded the Nobel Peace Prize. She is the first African woman to receive this famous award.

Energy Resources

Every day you use plants and animals from the dinosaur age—in the form of energy resources. These resources power vehicles, produce heat, and generate electricity. They are some of our most important and valuable natural resources.

Nonrenewable Energy Resources

Most of the energy we use comes from **fossil fuels**, nonrenewable resources that formed from the remains of ancient plants and animals. The most important fossil fuels are coal, petroleum, and natural gas.

Coal has long been a reliable energy source for heat. However, burning coal causes some problems. It pollutes the air and can harm the land. For these reasons, people have used coal less as other fuel options became available.

FOCUS ON READING

In the second sentence on this page, what cause does the word *because* signal? What is the effect of this cause?

Today we use coal mainly to create electricity at power plants, not to heat single buildings. Because coal is plentiful, people are looking for cleaner ways to burn it.

Petroleum, or oil, is a dark liquid used to make fuels and other products. When first removed from the ground, petroleum is called crude oil. This oil is shipped or piped to refineries, factories that process the crude oil to make products. Fuels made from oil include gasoline, diesel fuel, and jet fuel. Oil is also used to make petrochemicals, which are processed to make products such as plastics and cosmetics.

As with coal, burning oil-based fuels can pollute the air and land. In addition, oil spills can harm wildlife. Because we are so dependent on oil for energy, however, it is an extremely valuable resource.

The cleanest-burning fossil fuel is natural gas. We use it mainly for heating and cooking. For example, your kitchen stove may use natural gas. Some vehicles run on natural gas as well. These vehicles cause less pollution than those that run on gasoline.

Renewable Energy Resources

Unlike fossil fuels, renewable energy resources will not run out. They also are generally better for the environment. On the other hand, they are not available everywhere and can be costly.

The main alternative to fossil fuels is **hydroelectric power**—the production of electricity from waterpower. We obtain energy from moving water by damming rivers. The dams harness the power of moving water to generate electricity.

Hydroelectric power has both pros and cons. On the positive side, it produces power without polluting and lessens our use of fossil fuels. On the negative side, dams create lakes that replace existing resources, such as farmland, and disrupt wildlife habitats.

Another renewable energy source is wind. People have long used wind to power windmills. Today we use wind to power wind turbines, a type of modern windmill. At wind farms, hundreds of turbines create electricity in windy places.

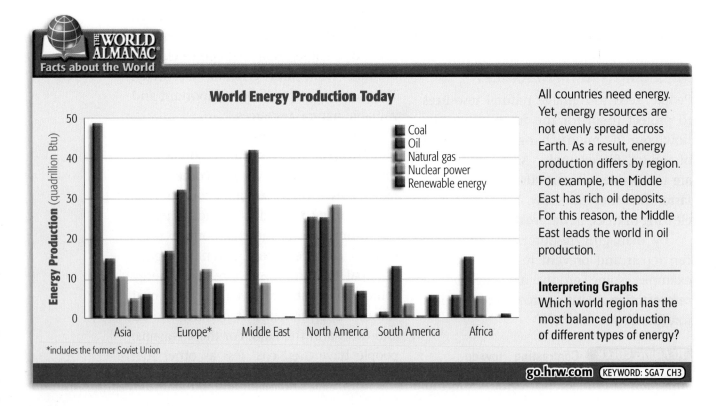

THE WORLD ALMANAC®
Facts about the World

World Energy Production Today

Energy Production (quadrillion Btu)

Legend:
- Coal
- Oil
- Natural gas
- Nuclear power
- Renewable energy

Regions: Asia, Europe*, Middle East, North America, South America, Africa

*includes the former Soviet Union

All countries need energy. Yet, energy resources are not evenly spread across Earth. As a result, energy production differs by region. For example, the Middle East has rich oil deposits. For this reason, the Middle East leads the world in oil production.

Interpreting Graphs
Which world region has the most balanced production of different types of energy?

go.hrw.com KEYWORD: SGA7 CH3

A third source of renewable energy is heat from the sun and Earth. We can use solar power, or power from the sun, to heat water or homes. Using special solar panels, we turn solar energy into electricity. We can also use geothermal energy, or heat from within Earth. Geothermal power plants use steam and hot water located within Earth to create electricity.

Nuclear Energy

A final energy source is nuclear energy. We obtain this energy by splitting atoms, small particles of matter. This process uses the metal uranium, so some people consider nuclear energy a nonrenewable resource. Nuclear power does not pollute the air, but it does produce dangerous wastes. These wastes must be stored for thousands of years before they are safe. In addition, an accident at a nuclear power plant can have terrible effects.

READING CHECK **Drawing Inferences** Why might people look for alternatives to fossil fuels?

Mineral Resources

Like energy resources, mineral resources can be quite valuable. These resources include metals, salt, rocks, and gemstones.

Minerals fulfill countless needs. Look around you to see a few. Your school building likely includes steel, made from iron. The outer walls might be granite or limestone. The window glass is made from quartz, a mineral in sand. From staples to jewelry to coins, metals are everywhere.

Minerals are nonrenewable, so we need to conserve them. Recycling items such as aluminum cans will make the supply of these valuable resources last longer.

READING CHECK **Categorizing** What are the major types of mineral resources?

From the Ground to the Air

The photo shows a bauxite mine. Bauxite is used to make aluminum. This metal is used in many products, such as jet planes. See how many other products made from aluminum you can name.

Resources and People

Natural resources vary from place to place. The resources available in a region can shape life and wealth for the people there.

Resources and Daily Life

The natural resources available to people affect their lifestyles and needs. In the United States we have many different kinds of natural resources. We can choose among many different ways to dress, eat, live, travel, and entertain ourselves. People in places with fewer natural resources will likely have fewer choices and different needs than Americans.

For example, people who live in remote rain forests depend on forest resources for most of their needs. These people may craft containers by weaving plant fibers together. They may make canoes by hollowing out tree trunks. Instead of being concerned about money, they might be more concerned about food.

Products from Petroleum

This Ohio family shows some common products made from petroleum, or oil.

ANALYZING VISUALS What petroleum-based products can you identify in this photo?

Resources and Wealth

The availability of natural resources affects countries' economies as well. For example, the many natural resources available in the United States have helped it become one of the world's wealthiest countries. In contrast, countries with few natural resources often have weak economies.

Some countries have one or two valuable resources but few others. For example, Saudi Arabia is rich in oil but lacks water for growing food. As a result, Saudi Arabia must use its oil profits to import food.

READING CHECK Identifying Cause and Effect How can having few natural resources affect life and wealth in a region or country?

SUMMARY AND PREVIEW You can see that Earth's natural resources have many uses. Important natural resources include air, water, soils, forests, fuels, and minerals. In the next chapter you will read about the world's people and cultures.

go.hrw.com
Online Quiz
KEYWORD: SGA7 HP3

Section 4 Assessment

Reviewing Key Ideas, Terms, and Places

1. **a. Define** What are **renewable resources** and **nonrenewable resources**?
   b. Explain Why is it important for people to manage Earth's natural resources?
   c. Develop What are some things you can do to help manage and conserve natural resources?
2. **a. Define** What are **fossil fuels**, and why are they significant?
   b. Summarize What are three examples of renewable energy resources?
   c. Predict How do you think life might change as we begin to run out of petroleum?
3. **a. Recall** What are the main types of mineral resources?
   b. Analyze What are some products that we get from mineral resources?

4. **a. Describe** How do resources affect people?
   b. Make Inferences How might a country with only one valuable resource develop its economy?

Critical Thinking

5. **Categorizing** Draw a chart like this one. Use your notes to identify and evaluate each energy resource.

Fossil Fuels	Renewable Energy	Nuclear Energy
Pros	Pros	Pros
Cons	Cons	Cons

FOCUS ON VIEWING

6. **Noting Details about Natural Resources** What natural resources does the place you chose have? Note ways to refer to some of these resources (or the lack of them) in your weather report.

from
The River

by Gary Paulsen

About the Reading *In the novel* The River, *a teenager named Brian has already proven his ability to survive in the wilderness. On this trip into the wilderness, he is accompanied by a man who wants to learn survival skills from him. With only a pocket knife and a transistor radio as tools, the two men meet challenges that at first appear too difficult to overcome. In the following passage from the novel, the men have just arrived in the wilderness.*

AS YOU READ Notice how Brian uses his senses to predict how some natural resources can help him survive.

He didn't just hear birds singing, not just a background sound of birds, but each bird. He listened to each bird. Located it, knew where it was by the sound, listened for the sound of alarm. He didn't just see clouds, but light clouds, scout clouds that came before the heavier clouds that could mean rain and maybe wind. ❶ The clouds were coming out of the northwest, and that meant that weather would come with them. Not could, but would. There would be rain. Tonight, late, there would be rain.

His eyes swept the clearing. . . There was a stump there that probably held grubs; hardwood there for a bow, and willows there for arrows; a game trail, . . . porcupines, raccoons, bear, wolves, moose, skunk would be moving on the trail and into the clearing. ❷ He flared his nostrils, smelled the air, pulled the air along the sides of his tongue in a hissing sound and tasted it, but there was nothing. Just summer smells. The tang of pines, soft air, some mustiness from rotting vegetation. No animals. ❸

GUIDED READING

WORD HELP

grubs soft, thick wormlike forms of insects
flared widened
tang sharp, biting smell
mustiness damp, stale smell

❶ Scout clouds are clouds that appear to be searching for other clouds to come.

❷ Brian notes that the stump likely holds grubs. He can eat the grubs for food.

❸ Brian does not smell any animals nearby.
Why might Brian want to know if animals are around?

Connecting Literature to Geography

1. **Predicting** Brian observes the clouds and can tell from their appearance and movement that rain is coming. What might be some ways that he can use his environment to prepare for the rain?

2. **Finding Main Ideas** The environment provides many resources that we can use, from wood to plants to animals. What resources does Brian identify around him that he can use to survive?

Social Studies Skills

Chart and Graph	Critical Thinking	Geography	Study

Analyzing a Bar Graph

Learn

Bar graphs are drawings that use bars to show data in a clear, visual format. Use these guidelines to analyze bar graphs.

- Read the title to identify the graph's subject and purpose.

- Read the graph's other labels. Note what the graph is measuring and the units of measurement being used. For example, this bar graph is measuring precipitation by climate. The unit of measurement is inches. If the graph uses colors, note their purpose.

- Analyze and compare the data. As you do, note any increases or decreases and look for trends or changes over time.

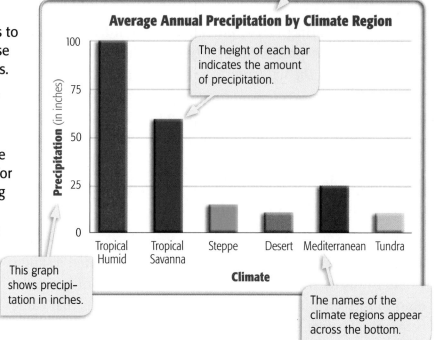

This bar graph compares the average annual precipitation of six climate regions.

The height of each bar indicates the amount of precipitation.

This graph shows precipitation in inches.

The names of the climate regions appear across the bottom.

Practice

❶ On the bar graph above, which climate region has the highest average annual precipitation?

❷ Which two climate regions have about the same amount?

❸ Which climate region receives an average of between 50 and 75 inches of precipitation each year?

Apply

Examine the World Energy Production Today bar graph in Section 4. Then use the graph to answer the following questions.

1. Which region produces the most oil?

2. Which three regions produce little or no nuclear power?

3. Based on the graph, what type of energy resource do most Asian countries likely use?

Chapter Review

Geography's Impact
video series
Review the video to answer the closing question:
How are climate and weather different, and how does the influence they have differ?

Visual Summary

Use the visual summary below to help you review the main ideas of the chapter.

QUICK FACTS

Earth has a wide range of climates, which we identify by precipitation, temperature, and native plant life.

Plants, animals, and the environment are interconnected and affect one another in many ways.

Earth's valuable natural resources, such as air, water, forests, and minerals, have many uses and affect people's lives.

Reviewing Vocabulary, Terms, and Places

Unscramble each group of letters below to spell a term that matches the given definition.

1. **usumh**—decayed plant or animal matter
2. **tahrewe**—changes or conditions in the air at a certain time and place
3. **netorietfaosr**—planting trees where forests were
4. **neticxt**—completely died out
5. **estpep**—semidry grassland or prairie
6. **sifeticatorined**—spread of desertlike conditions
7. **laitemc**—an area's weather patterns over a long period of time
8. **arsmofrtpe**—permanently frozen layers of soil
9. **snonomo**—winds that change direction with the seasons and create wet and dry periods
10. **vansanas**—areas of tall grasses and scattered shrubs and trees

Comprehension and Critical Thinking

SECTION 1 *(Pages 50–54)*

11. **a. Identify** What five factors affect climate?

    b. Analyze Is average annual precipitation an example of weather or climate?

    c. Evaluate Of the five factors that affect climate, which one do you think is the most important? Why?

SECTION 2 *(Pages 55–61)*

12. **a. Recall** What are the five major climate zones?

    b. Explain How does latitude relate to climate?

    c. Elaborate Why do you think the study of climate is important in geography?

SECTION 3 *(Pages 62–65)*

13. **a. Define** What is an ecosystem, and why does it exist in a fragile balance?

SECTION 3 *(continued)*

b. Explain Why are plants an important part of the environment?

c. Predict What might be some results of desertification?

SECTION 4 *(Pages 68–72)*

14. a. Define What are minerals?

b. Contrast How do nonrenewable resources and renewable resources differ?

c. Elaborate How might a scarcity of natural resources affect life in a region?

Using the Internet

15. Activity: Experiencing Extremes Could you live in a place where for part of the year it is always dark and temperatures plummet to –104°F? What if you had to live in a place where it is always wet and stormy? Enter the activity keyword to learn more about some of the world's extreme climates. Then create a poster that describes some of those climates and the people, animals, and plants that live in them.

FOCUS ON READING AND VIEWING

Understanding Cause and Effect *Answer the following questions about causes and effects.*

16. What causes desertification?

17. What are the effects of abundant natural resources on a country's economy?

Presenting and Viewing a Weather Report *Use your weather report notes to complete the activity below.*

18. Select a place and a season. Then write a script for a weather report for that place during that season. Describe the current weather and predict the upcoming weather. During your presentation, use a professional, friendly tone of voice and make frequent eye contact with your audience. Then view your classmates' weather reports. Be prepared to give feedback on the content and their presentation techniques.

Social Studies Skills

Analyzing a Bar Graph *Examine the bar graph titled Average Annual Precipitation by Climate Region in the Social Studies Skills for this chapter. Then use the bar graph to answer the following questions.*

19. Which climate region receives an average of 100 inches of precipitation a year?

20. Which climate region receives an average of 25 inches of precipitation a year?

21. What is the difference in average annual precipitation between tropical humid climates and Mediterranean climates?

Map Activity

22. Prevailing Winds On a separate sheet of paper, match the letters on the map with their correct labels.

| equator | South Pole | westerly |
| North Pole | trade wind | |

DIRECTIONS: Read questions 1 through 7 and write the letter of the best response. Then read question 8 and write your own well-constructed response.

1 The cold winds that flow away from the North and South poles are the

A doldrums.

B polar easterlies.

C trade winds.

D westerlies.

2 Which climate zone occurs only in the upper latitudes?

A highland

B temperate

C tropical

D polar

3 Where are the most diverse habitats on Earth found?

A steppe

B tropical rain forest

C tropical savanna

D tundra

4 What is the cleanest burning fossil fuel?

A coal

B natural gas

C oil

D petroleum

5 Which renewable energy source uses the heat of Earth's interior to generate power?

A geothermal energy

B hydroelectric energy

C nuclear energy

D solar energy

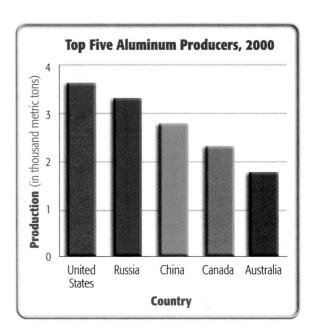

6 Based on the graph above, which country produced about 2,750 metric tons of aluminum in 2000?

A Australia

B China

C Russia

D United States

7 Which of the following form over tropical waters and are Earth's largest and most destructive storms?

A blizzards

B hurricanes

C thunderstorms

D tornadoes

8 Extended Response Forces such as the sun, latitude, wind, and water shape climate. Examine the World Climate Regions map in Section 2. Describe two climate patterns that you see on the map and explain how various forces combine to create the two patterns.

CHAPTER 4

The World's People

What You Will Learn...

In this chapter you will learn what culture is and how it changes over time. You will also study population and the different types of governments and economic systems used around the world. Finally, you will discover how global connections are bringing the world's people closer together.

SECTION 1
Culture **80**

SECTION 2
Population **86**

SECTION 3
Government and Economy **91**

SECTION 4
Global Connections.................... **97**

FOCUS ON READING AND WRITING

Understanding Main Ideas A main idea is the central idea around which a paragraph or passage is organized. As you read, ask yourself what each paragraph is about. Look for a sentence or two that summarizes the main point of the entire paragraph. **See the lesson, Understanding Main Ideas, on page 109.**

Creating a Poster Think of some great posters you have seen—at the movies, in bus stations, or in the halls of your school. They likely all had a colorful image that captured your attention and a few words that explained the main idea. Read this chapter about the world's people. Then create a poster that includes words and images that summarize the chapter's main ideas.

Culture Thousands of different cultures make up our world. Clothing, language, and music are just some parts of culture.

ANALYSIS SKILL   **ANALYZING VISUALS**

Many of the world's people come together every four years to compete in the Olympics.

What indicates that some of the people in this photo are from different parts of the world?

HOLT

Geography's Impact
video series
Watch the video to learn the impact of culture.

Population Geographers study human populations like this one in India to learn where and why people live in certain places.

Global Connections
Technology allows people in remote places around the world to communicate.

Culture

If YOU lived there...

You live in New York City, and your young cousin from out of state has come to visit. As you take her on a tour of the city, you point out the different cultural neighborhoods, like Chinatown, Little Italy, Spanish Harlem, and Koreatown. Your cousin isn't quite sure what culture means or why these neighborhoods are so different.

How can you explain what culture is?

BUILDING BACKGROUND For hundreds of years, immigrants from around the world have moved to the United States to make a new home here. They have brought with them all the things that make up culture—language, religion, beliefs, traditions, and more. As a result, the United States has one of the most diverse cultures in the world.

What Is Culture?

If you traveled around the world, you would experience many different sights and sounds. You would probably hear unique music, eat a variety of foods, listen to different languages, see distinctive landscapes, and learn new customs. You would see and take part in the variety of cultures that exist in our world.

A Way of Life

What exactly is culture? **Culture** is the set of beliefs, values, and practices that a group of people has in common. Culture includes many aspects of life, such as language and religion, that we may share with people around us. Everything in your day-to-day life is part of your culture, from the clothes you wear to the music you hear to the foods you eat.

On your world travels, you might notice that all societies share certain cultural features. All people have some kind of government, educate their children in some way, and create some type of art or music. However, not all societies practice their culture in the same way. For example, in Japan the school year begins in the spring, and students wear school uniforms. In the United States, however, the school year begins in the late

What You Will Learn...

Main Ideas

1. Culture is the set of beliefs, goals, and practices that a group of people share.
2. The world includes many different culture groups.
3. New ideas and events lead to changes in culture.

The Big Idea

Culture, a group's shared practices and beliefs, differs from group to group and changes over time.

Key Terms

culture, *p. 80*
culture trait, *p. 81*
culture region, *p. 82*
ethnic group, *p. 83*
cultural diversity, *p. 83*
cultural diffusion, *p. 85*

TAKING NOTES As you read, take notes on culture. Use a web diagram like the one below to organize your notes.

Culture Traits

These students in Japan and Kenya have some culture traits in common, like eating lunch at school. Other culture traits are different.

ANALYZING VISUALS What culture traits do these students share? Which are different?

summer, and most schools do not require uniforms. Differences like these are what make each culture unique.

Culture Traits

Cultural features like starting the school year in the spring or wearing uniforms are types of culture traits. A **culture trait** is an activity or behavior in which people often take part. The language you speak and the sports you play are some of your culture traits. Sometimes a culture trait is shared by people around the world. For example, all around the globe people participate in the game of soccer. In places as different as Germany, Nigeria, and Saudi Arabia, many people enjoy playing and watching soccer.

While some culture traits are shared around the world, others change from place to place. One example of this is how people around the world eat. In China most people use chopsticks to eat their food. In Europe, however, people use forks and spoons. In Ethiopia, many people use bread or their fingers to scoop their food.

Development of Culture

How do cultures develop? Culture traits are often learned or passed down from one generation to the next. Most culture traits develop within families as traditions, foods, or holiday customs are handed down over the years. Laws and moral codes are also passed down within societies. Many laws in the United States, for example, can be traced back to England in the 1600s and were brought by colonists to America.

Cultures also develop as people learn new culture traits. Immigrants who move to a new country, for example, might learn to speak the language or eat the foods of their adopted country.

Other factors, such as history and the environment, also affect how cultures develop. For example, historical events changed the language and religion of much of Central and South America. In the 1500s, when the Spanish conquered the region, they introduced their language and Roman Catholic faith. The environment in which we live can also shape

FOCUS ON READING
What is the main idea of this paragraph?

For example, the desert environment of Africa's Sahara influences the way people who live there earn a living. Rather than grow crops, they herd animals that have adapted to the harsh environment. As you can see, history and the environment affect how cultures develop.

> **READING CHECK** **Finding Main Ideas** What practices and customs make up culture?

Culture Groups

Earth is home to thousands of different cultures. People who share similar culture traits are members of the same culture group. Culture groups can be based on a variety of factors, such as age, language, or religion. American teenagers, for example, can be said to form a culture group based on location and age. They share similar tastes in music, clothing, and sports.

Culture Regions

When we refer to culture groups, we are speaking of people who share a common culture. At other times, however, we need to refer to the area, or region, where the culture group is found. A **culture region** is an area in which people have many shared culture traits.

In a specific culture region, people share certain culture traits, such as religious beliefs, language, or lifestyle. One well-known culture region is the Arab world. As you can see at right, an Arab culture region spreads across Southwest Asia and North Africa. In this region, most people write and speak Arabic and are Muslim. They also share other traits, such as foods, music, styles of clothing, and architecture.

Occasionally, a single culture region dominates an entire country. In Japan, for example, one primary culture dominates the country. Nearly everyone in Japan speaks the same language and follows the same practices. Many Japanese bow to their elders as a sign of respect and remove their shoes when they enter a home.

A single country may also include more than one culture region within its borders. Mexico is one of many countries that is made up of different culture regions. People in northern Mexico and southern Mexico, for example, have different culture traits. The culture of northern Mexico tends to be more modern, while traditional culture remains strong in southern Mexico.

A culture region may also stretch across country borders. As you have already learned, an Arab culture region dominates much of Southwest Asia and North Africa. Another example is the Kurdish culture region, home to the Kurds, a people that live throughout Turkey, Iran, and Iraq.

Arab Culture Region

Culture regions are based on shared culture traits. Southwest Asia and North Africa make up an Arab culture region based on ethnic heritage, a common language, and religion. Most people in this region are Arab, speak and write Arabic, and practice Islam.

EUROPE

ASIA

AFRICA

Cultural Diversity

As you just learned, countries may contain several culture regions within their borders. Often, these culture regions are based on ethnic groups. An **ethnic group** is a group of people who share a common culture and ancestry. Members of ethnic groups often share certain culture traits such as religion, language, and even special foods.

Some countries are home to a variety of ethnic groups. For example, more than 100 different ethnic groups live in the East African country of Tanzania. Countries with many ethnic groups are culturally diverse. **Cultural diversity** is the state of having a variety of cultures in the same area. While cultural diversity creates an interesting mix of ideas, behaviors, and practices, it can also lead to conflict.

In some countries, ethnic groups have been in conflict. In Canada, for example, some French Canadians want to separate from the rest of Canada to preserve their language and culture. In the 1990s ethnic conflict in the African country of Rwanda led to extreme violence and bloodshed.

Although ethnic groups have clashed in some culturally diverse countries, they have cooperated in others. In the United States, for example, many different ethnic groups live side by side. Cities and towns often celebrate their ethnic heritage with festivals and parades, like the Saint Patrick's Day Parade in Boston or Philadelphia's Puerto Rican Festival.

READING CHECK **Making Inferences** Why might cultural diversity cause conflict?

Many people share Arab culture traits. An Algerian boy, above, and Palestinian girls, at left, share the same language and religion.

ANALYZING VISUALS What culture traits do you see in the photos?

83

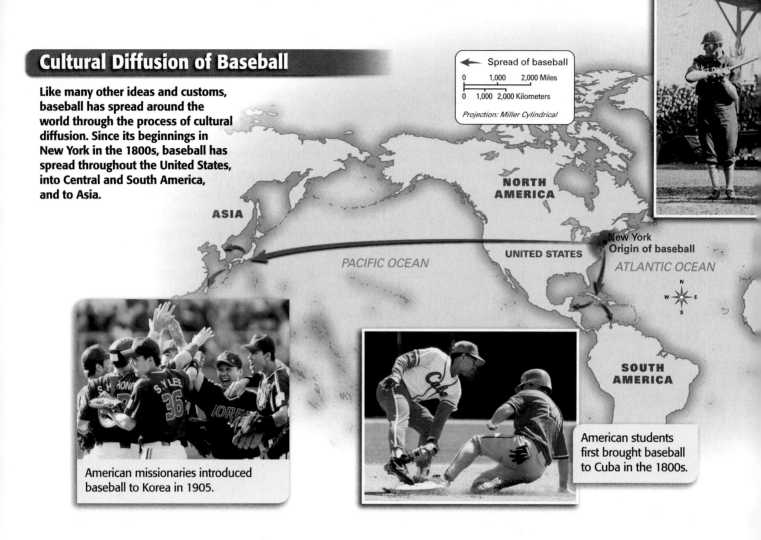

Cultural Diffusion of Baseball

Like many other ideas and customs, baseball has spread around the world through the process of cultural diffusion. Since its beginnings in New York in the 1800s, baseball has spread throughout the United States, into Central and South America, and to Asia.

Spread of baseball

0   1,000   2,000 Miles

0   1,000 2,000 Kilometers

Projection: Miller Cylindrical

ASIA

NORTH AMERICA

PACIFIC OCEAN

UNITED STATES

New York Origin of baseball

ATLANTIC OCEAN

SOUTH AMERICA

American missionaries introduced baseball to Korea in 1905.

American students first brought baseball to Cuba in the 1800s.

Changes in Culture

You've read books or seen movies set in the time of the Civil War or in the Wild West of the late 1800s. Think about how our culture has changed since then. Clothing, food, music—all have changed drastically. When we study cultural change, we try to find out what caused the changes and how those changes spread from place to place.

How Cultures Change

Cultures change constantly. Some changes happen rapidly, while others take many years. What causes cultures to change? **Innovation** and contact with other people are two key causes of cultural change.

New ideas often bring about cultural changes. For example, when Alexander Graham Bell invented the telephone, it changed how people communicate with each other. Other innovations, such as motion pictures, changed how people spend their free time. More recently, the creation of the Internet dramatically altered the way people find information, communicate, and shop.

Cultures also change as societies come into contact with each other. For example, when the Spanish arrived in the Americas, they introduced firearms and horses to the region, changing the lifestyle of some Native American groups. At the same time, the Spaniards learned about new foods like potatoes and chocolate. These foods then became an important part of Europeans' diet. The Chinese had a similar influence on Korea and Japan, where they introduced Buddhism and written language.

ACADEMIC VOCABULARY

innovation a new idea or way of doing something

Organized baseball began in New York around 1845 and quickly spread around the world.

Where did baseball begin, and to what parts of the world did it eventually spread?

How Ideas Spread

You have probably noticed that a new slang word might spread from teenager to teenager and state to state. In the same way, clothing styles from New York or Paris might become popular all over the world. More serious cultural traits spread as well. Religious beliefs or ideas about government may spread from place to place. The spread of culture traits from one region to another is called **cultural diffusion**.

Cultural diffusion often occurs when people move from one place to another. For example, when Europeans settled in the Americas, they brought their culture along with them. As a result, English, French, Spanish, and Portuguese are all spoken in the Americas. American culture also spread as pioneers moved west, taking with them their form of government, religious beliefs, and customs.

Cultural diffusion also takes place as new ideas spread from place to place. As you can see on the map above, the game of baseball first began in New York, then spread throughout the United States. As more and more people learned the game, it spread even faster and farther. Baseball eventually spread around the world. Wearing blue jeans became part of our culture in a similar way. Blue jeans originated in the American West in the mid-1800s. They gradually became popular all over the country and the world.

READING CHECK **Finding Main Ideas** How do cultures change over time?

SUMMARY AND PREVIEW In this section you learned about the role that culture plays in our lives and how our cultures change. Next, you will learn about human populations and how we keep track of Earth's changing population.

Section 1 Assessment

go.hrw.com
Online Quiz
KEYWORD: SGA7 HP4

Reviewing Key Ideas, Terms, and Places

1. **a. Define** What is **culture**?
   b. Analyze What influences the development of culture?
   c. Elaborate How might the world be different if we all shared the same culture?
2. **a. Identify** What are the different types of **culture regions**?
   b. Analyze How does **cultural diversity** affect societies?
3. **a. Describe** How does **cultural diffusion** take place?
   b. Make Inferences How can the spread of new ideas lead to cultural change?
   c. Evaluate Do you think that cultural diffusion has a positive or a negative effect? Explain your answer.

Critical Thinking

4. **Finding Main Ideas** Using your notes and a chart like the one here, explain the main idea of each aspect of culture in your own words.

Culture Traits	Culture Groups	Cultural Change

FOCUS ON WRITING

5. **Writing about Culture** What key words about culture can you include on your poster? What images might you include? Jot down your ideas in your notebook.

Population

If YOU lived there...

You live in Mexico City, one of the largest and most crowded cities in the world. You realize just how crowded it is whenever you ride the subway at rush hour! You love the excitement of living in a big city. There is always something interesting to do. At the same time, the city has a lot of crime. Heavy traffic pollutes the air.

What do you like and dislike about living in a large city?

BUILDING BACKGROUND An important part of geographers' work is the study of human populations. Many geographers are interested in where people live, how many people live there, and what effects those people have on resources and the environment.

Population Patterns

How many people live in your community? Do you live in a small town, a huge city, or somewhere in between? Your community's **population**, or the total number of people in a given area, determines a great deal about the place in which you live. Population influences the variety of businesses, the types of transportation, and the number of schools in your community.

Because population has a huge impact on our lives, it is an important part of geography. Geographers who study human populations are particularly interested in patterns that emerge over time. They study such information as how many people live in an area, why people live where they do, and how populations change. Population patterns like these can tell us much about our world.

Population Density

Some places on Earth are crowded with people. Others are almost empty. One statistic geographers use to examine populations is **population density**, a measure of the number of people living in an area. Population density is expressed as persons per square mile or square kilometer.

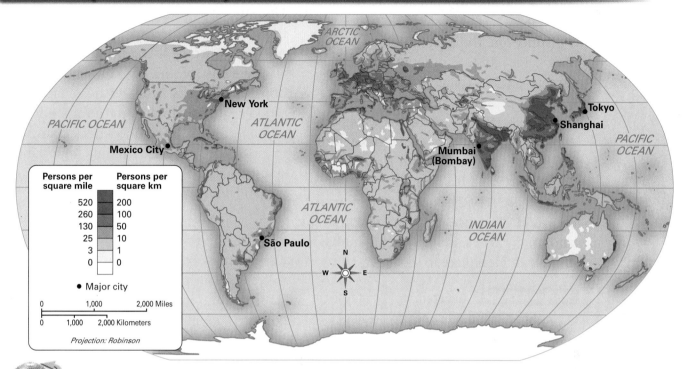

map zone

Geography Skills

Location While low population densities are common throughout much of the world, South and East Asia are two of the world's most densely populated regions.

1. **Identify** Which continent is the most densely populated? Which is the least densely populated?
2. **Making Inferences** Why might the population density of far North America be so low?

go.hrw.com (KEYWORD: SGA7 CH4)

Population density provides us with important information about a place. The more people per square mile in a region, the more crowded, or dense, it is. Japan, for example, has a population density of 873 people per square mile (340 per square km). That is a high population density. In many parts of Japan, people are crowded together in large cities, and space is very limited. In contrast, Australia has a very low population density. Only 7 people per square mile (3 per square km) live there. Australia has many wide-open spaces with very few people.

How do you think population density affects life in a particular place? In places with high population densities, the land is often expensive, roads are crowded, and buildings tend to be taller. On the other hand, places with low population densities tend to have more open spaces, less traffic, and more available land.

Where People Live

Can you tell where most of the world's people live by examining the population density map above? The reds and purples on the map indicate areas of very high population density, while the light yellow areas indicate sparse populations. When an area is thinly populated, it is often because the land does not provide a very good life. These areas may have rugged mountains or harsh deserts where people cannot grow crops. Some areas may be frozen all year long, making survival there very difficult.

For these reasons, very few people live in parts of far North America, Greenland, northern Asia, and Australia.

Notice on the map that some areas have large clusters of population. Such clusters can be found in East and South Asia, Europe, and eastern North America. Fertile soil, reliable sources of water, and a good agricultural climate make these good regions for settlement. For example, the North China Plain in East Asia is one of the most densely populated regions in the world. The area's plentiful agricultural land, many rivers, and mild climate have made it an ideal place to settle.

READING CHECK **Generalizing** What types of information can population density provide?

CONNECTING TO Math

Calculating Population Density

Population density measures the number of people living in an area. To calculate population density, divide a place's total population by its area in square miles (or square kilometers). For example, if your city has a population of 100,000 people and an area of 100 square miles, you would divide 100,000 by 100. This would give you a population density of 1,000 people per square mile (100,000 ÷ 100 = 1,000).

Analyzing If a city had a population of 615,000 and a total land area of 250 square miles, what would its population density be?

City	Population	Total Area (square miles)	Population Density (people per square mile)
Adelaide, Australia	1,032,585	336	3,073
Lima, Peru	8,043,521	1,029	7,816
Nairobi, Kenya	2,143,254	266	8,057

Population Change

The study of population is much more important than you might realize. The number of people living in an area affects all elements of life—the availability of housing and jobs, whether hospitals and schools open or close, even the amount of available food. Geographers track changes in populations by examining important statistics, studying the movement of people, and analyzing population trends.

Tracking Population Changes

Geographers examine three key statistics to learn about population changes. These statistics are important for studying a country's population over time.

Three key statistics—birthrate, death rate, and the rate of natural increase—track changes in population. Births add to a population. Deaths subtract from it. The annual number of births per 1,000 people is called the **birthrate**. Similarly, the death rate is the annual number of deaths per 1,000 people. The birthrate minus the death rate equals the percentage of natural increase, or the rate at which a population is changing. For example, Japan has a rate of natural increase of .07%. This means it has slightly more births than deaths and a very slight population increase.

Population growth rates differ from one place to another. In some countries, populations are growing very slowly or even shrinking. Many countries in Europe and North America have very low rates of natural increase. In Russia, for example, the birthrate is about 9.6 and the death rate is 15.2. The result is a negative rate of natural increase and a shrinking population.

In most countries around the world, however, populations are growing. Mali, for example, has a rate of natural increase of almost 3 percent. While that may sound

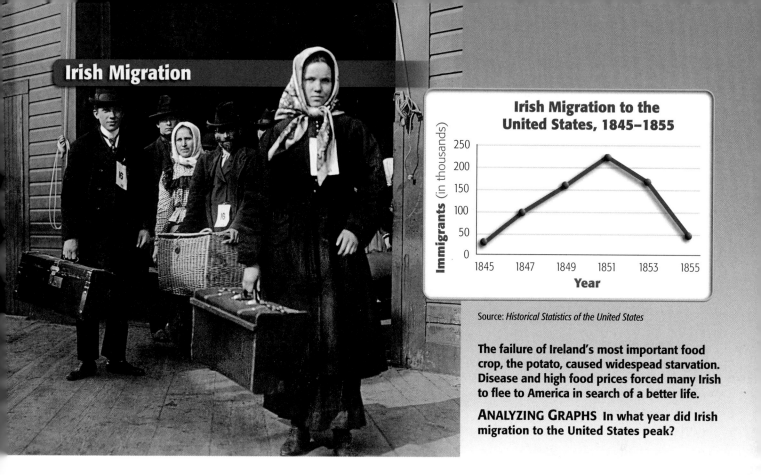

Irish Migration

Irish Migration to the United States, 1845–1855

Source: *Historical Statistics of the United States*

The failure of Ireland's most important food crop, the potato, caused widespread starvation. Disease and high food prices forced many Irish to flee to America in search of a better life.

ANALYZING GRAPHS In what year did Irish migration to the United States peak?

small, it means that Mali's population is expected to double in only 23 years! High population growth rates can pose some challenges, as governments try to provide enough jobs, education, and medical care for their rapidly growing populations.

Migration

A common cause of population change is migration. **Migration** is the process of moving from one place to live in another. As one country loses citizens as a result of migration, its population can decline. At the same time, another country may gain population as people settle there.

People migrate for many reasons. Some factors push people to leave their country, while other factors pull, or attract, people to new countries. Warfare, a lack of jobs, or a lack of good farmland are common push factors. For example, during the Irish potato famine of the mid-1800s, poverty and disease forced some 1.5 million people

to leave Ireland. Opportunities for a better life often pull people to new countries. For example, in the 1800s and early 1900s thousands of British citizens migrated to Australia in search of cheap land.

World Population Trends

In the last 200 years Earth's population has exploded. For thousands of years world population growth was low and relatively steady. About 2,000 years ago, the world had some 300 million people. By 1800 there were almost 1 billion people. Since 1800, better health care and improved food production have supported tremendous population growth. In 1999 the world's population passed 6 billion people.

Population trends are an important part of the study of the world's people. Two important population trends are clear today. The first trend indicates that the population growth in some of the more industrialized nations has begun to slow.

FOCUS ON READING
What is the main idea of this paragraph? What facts are used to support that idea?

World Population Growth

Advances in food production and health care have dramatically lowered death rates. As a result, the global population has seen incredible growth over the last 200 years.

ANALYZING GRAPHS By how much did the world's population increase between 1800 and 2000?

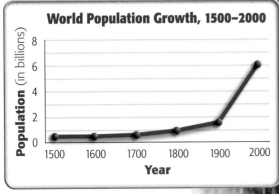

World Population Growth, 1500–2000

Source: *Atlas of World Population History*

For example, Germany and France have low rates of natural increase. A second trend indicates that less industrialized countries, like Nigeria and Bangladesh, often have high growth rates. These trends affect a country's workforce and government aid.

READING CHECK **Summarizing** What population statistics do geographers study? Why?

SUMMARY AND PREVIEW In this section you have learned where people live, how crowded places are, and how population affects our world. Geographers study past and present population patterns in order to plan for the future. In the next section, you will learn how governments and economies affect people on Earth.

Section 2 Assessment

Reviewing Key Ideas, Terms, and Places

1. **a. Identify** What regions of the world have the highest levels of **population density**?
   b. Draw Conclusions What information can be learned by studying population density?
   c. Evaluate Would you prefer to live in a region with a dense or a sparse population? Why?
2. **a. Describe** What is natural increase? What can it tell us about a country?
   b. Analyze What effect does **migration** have on human populations?
   c. Predict What patterns do you think world population might have in the future?

Critical Thinking

3. **Summarizing** Draw a graphic organizer like the one here. Use your notes to write a sentence that summarizes each aspect of the study of population.

Population Patterns

Population Change

FOCUS ON WRITING

4. **Discussing Population** What effect does population have on our world? Write down some words and phrases that you might use on your poster to explain the importance of population.

Government and Economy

If YOU lived there...

You live in Raleigh, North Carolina. Your class at school is planning a presentation about life in the United States for a group of visitors from Japan. Your teacher wants you to discuss government and economics in the United States. As you prepare for your speech, you wonder what you should say.

How do government and economics affect your life?

BUILDING BACKGROUND Although you probably don't think about them every day, your country's government and economy have a big influence on your life. That is true in every country in every part of the world. Governments and economic systems affect everything from a person's rights to the type of job he or she has.

Governments of the World

Can you imagine what life would be like if there were no rules? Without ways to establish order and ensure justice, life would be chaotic. That explains why societies have governments. Our governments make and enforce laws, regulate business and trade, and provide aid to people. Governments help shape the culture and economy of a country as well as the daily lives of the people who live there.

Democratic Governments

Many countries—including the United States, Canada, and Mexico—have democratic governments. A **democracy** is a form of government in which the people elect leaders and rule by majority. In most democratic countries, citizens are free to choose representatives to make and enforce the laws. Voters in the United States, for example, elect members of Congress, who make the laws, and the president, who enforces those laws.

What You Will Learn...

Main Ideas

1. The governments of the world include democracy, monarchy, dictatorship, and communism.
2. Different economic activities and systems exist throughout the world.
3. Geographers group the countries of the world based on their level of economic development.

The Big Idea

The world's countries have different governments and levels of economic development.

Key Terms

democracy, *p. 91*
communism, *p. 92*
market economy, *p. 94*
command economy, *p. 94*
gross domestic product (GDP), *p. 95*
developed countries, *p. 95*
developing countries, *p. 95*

TAKING NOTES As you read, use a chart like this one to take notes on the different types of governments and economies.

Government	Economy

Governments of the World

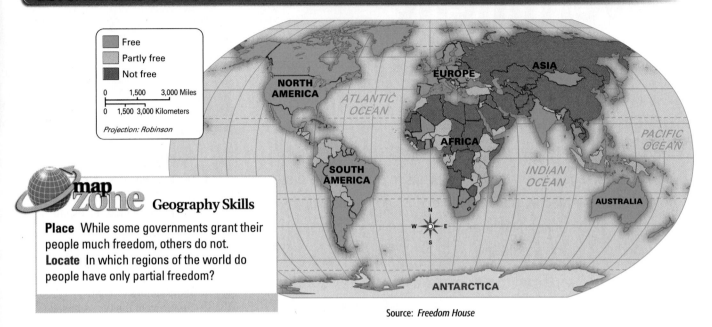

■	Free
■	Partly free
■	Not free

0   1,500   3,000 Miles

0   1,500 3,000 Kilometers

Projection: Robinson

map zone Geography Skills

Place While some governments grant their people much freedom, others do not.
Locate In which regions of the world do people have only partial freedom?

Source: *Freedom House*

FOCUS ON READING

Main ideas are not always stated in the first sentence. Which sentence in this paragraph states the main idea?

Most democratic governments in the world work to protect the freedoms and rights of their people, such as the freedom of speech and the freedom of religion. Other democracies, however, restrict the rights and freedoms of their people. Not all democratic governments in the world are completely free.

Other Types of Government

Not all of the world's countries, however, are democracies. Several other types of government are found in the world today, including monarchies, dictatorships, and Communist states.

Monarchy is one of the oldest types of government in the world. A monarchy is ruled directly by a king or queen, the head of a royal family. Saudi Arabia is an example of a monarchy. The Saudi king has executive, legislative, and judicial powers. In some monarchies, power is in the hands of just one person. As a result, the people have little say in their government. Other monarchies, however, like Norway and Spain, use many democratic practices.

Dictatorship is a type of government in which a single, powerful ruler has total control. This leader, called a dictator, often rules by force. Iraq under Saddam Hussein was an example of a dictatorship. People who live under a dictatorship are not free. They have few rights and no say in their own government.

Yet another form of government is communism. **Communism** is a political system in which the government owns all property and dominates all aspects of life in a country. Leaders of most Communist governments are not elected by citizens. Rather, they are chosen by the Communist Party or by Communist leaders. In most Communist states, like Cuba and North Korea, the government strictly controls the country's economy and the daily life of its people. As a result, people in Communist states often have restricted rights and very little freedom.

READING CHECK **Supporting a Point of View** Why might people prefer to live in a democracy as opposed to a dictatorship?

Economies of the World

One important function of government is to monitor a country's economy. The economy is a system that includes all of the activities that people and businesses do to earn a living. Countries today use a mix of different economic activities and systems.

Economic Activity

Every country has some level of economic activity. Economic activities are ways in which people make a living. Some people farm, others manufacture goods, while still others provide services, such as driving a taxi or designing skyscrapers. Geographers divide these economic activities into four different levels.

The first level of economic activity, the primary industry, uses natural resources or raw materials. People in these industries earn a living by providing raw materials to others. Farming, fishing, and mining are all examples of primary industries. These activities provide raw materials such as grain, seafood, and coal for others to use.

Secondary industries perform the next step. They use natural resources or raw materials to manufacture other products. Manufacturing is the process in which raw materials are changed into finished goods. For example, people who make furniture might take wood and make products such as tables, chairs, or desks. Automobile manufacturers use steel, plastic, glass, and rubber to put together trucks and cars.

In the third level of activity, or tertiary industry, goods and services are exchanged. People in tertiary industries sell the furniture, automobiles, or other products made in secondary industries. Other people, like health care workers or mechanics, provide services rather than goods. Teachers, store clerks, doctors, and TV personalities are all engaged in this level of economic activity.

Economic Activity

Primary Industry

Primary industries use natural resources to make money. Here a farmer sells milk from dairy cows to earn a living.

Secondary Industry

Secondary economic activities use raw materials to produce or manufacture something new. In this case, the milk from dairy cows is used to make cheese.

Tertiary Industry

Tertiary economic activities provide services to people and businesses. This grocer selling cheese in a market is involved in a tertiary activity.

Quaternary Industry

Quaternary industries process and distribute information. Skilled workers research and gather information. Here, inspectors examine and test the quality of cheese.

The highest level of economic activity, quaternary industry, involves the research and distribution of information. People making a living at this level work with information rather than goods, and often have specialized knowledge and skills. Architects, lawyers, and scientists all work in quaternary industries.

Economic Systems

Just as economic activities are organized into different types, so are our economic systems. Economic systems can be divided into three types: traditional, market, and command. Most countries today use a mix of these economic systems.

ACADEMIC VOCABULARY
traditional customary, time-honored

One economic system is a **traditional** economy, a system in which people grow their own food and make their own goods. Trade may take place through barter, or the exchange of goods without the use of money. Rural and remote communities often have a mostly traditional economy.

The most common economic system used around the world today is a market economy. A **market economy** is a system based on private ownership, free trade, and competition. Individuals and businesses are free to buy and sell what they wish. Prices are determined by the supply and demand for goods. This is sometimes called capitalism. The United States is one of many countries that use this system.

A third system is a **command economy**, a system in which the central government makes all economic decisions. The government decides what goods to produce, how much to produce, and what prices will be. While no country has a purely command economy, the economies of North Korea and Cuba are close to it. The Communist governments of these nations own and control most businesses.

READING CHECK **Summarizing** What economic systems are used in the world today?

THE WORLD ALMANAC **Facts about Countries** **A Developed and a Developing Country**

Australia	Afghanistan
Per Capita GDP (U.S. $): $30,700	Per Capita GDP (U.S. $): $800
Life Expectancy at Birth: 80.4	Life Expectancy at Birth: 42.9
Literacy Rate: 100%	Literacy Rate: 36%
Physicians Per 10,000 People: 25	Physicians Per 10,000 People: 1.9

Contrasting How does the quality of life in Afghanistan differ from that in Australia?

go.hrw.com KEYWORD: SGA7 CH4

Economic Development

Economic systems and activities affect a country's economic development, or the level of economic growth and quality of life. Geographers often group countries into two basic categories—developed and developing countries—based on their level of economic development.

Economic Indicators

Geographers use economic indicators, or measures of a country's wealth, to decide if a country is developed or developing. One such measure is gross domestic product. **Gross domestic product (GDP)** is the value of all goods and services produced within a country in a single year. Another indicator is a country's per capita GDP, or the total GDP divided by the number of people in a country. As you can see in the chart, per capita GDP allows us to compare incomes among countries. Other indicators include the level of industrialization and overall quality of life. In other words, we look at the types of industries and technology a country has, in addition to its level of health care and education.

Developed and Developing Countries

Many of the world's wealthiest and most powerful nations are **developed countries**, countries with strong economies and a high quality of life. Developed countries like Germany and the United States have a high per capita GDP and high levels of industrialization. Their health care and education systems are among the best in the world. Many people in developed countries have access to technology.

The world's poorer nations are known as **developing countries**, countries with less productive economies and a lower quality of life. Almost two-thirds of the people

in the world live in developing countries. These developing countries have a lower per capita GDP than developed countries. Most of their citizens work in farming or other primary industries. Although these countries typically have large cities, much of their population still lives in rural areas. People in developing countries usually have less access to health care or technology. Guatemala, Nigeria, and Afghanistan are all developing countries.

READING CHECK **Analyzing** What factors separate developed and developing countries?

SUMMARY AND PREVIEW The world's countries have different governments, economies, and levels of development. Next, you will learn how people are linked in a global community.

Section 3 Assessment

go.hrw.com
Online Quiz
KEYWORD: SGA7 HP4

Reviewing Key Ideas, Terms, and Places

1. **a. Identify** What are some different types of government?
   b. Evaluate Under which type of government would you most want to live? Why?
2. **a. Describe** What are the levels of economic activity?
   b. Evaluate Which economic system do you think is best? Explain your answer.
3. **a. Define** What is **gross domestic product**?
   b. Contrast In what ways do **developed countries** differ from **developing countries**?

Critical Thinking

4. **Categorizing** Draw a chart like the one here. Use the chart and your notes to identify the different governments, economies, and levels of economic development in the world today.

Types of Government	Economic Systems	Economic Development

FOCUS ON WRITING

5. **Thinking about Government and Economy** What kind of images and words might you use to present the main ideas behind the world's governments and economies?

Social Studies Skills

Chart and Graph   Critical Thinking   Geography   Study

Organizing Information

Learn

Remembering new information is easier if you organize it clearly. As you read and study, try to organize what you are learning. One way to do this is to create a graphic organizer. Follow these steps to create a graphic organizer as you read.

- Identify the main idea of the passage. Write the main idea in a circle at the top of your page.

- As you read, look for subtopics under the main idea. On your paper, draw a row of circles below the main idea, one for each subtopic. Write the subtopics in the circles.

- Below each subtopic, draw a big box. Look for facts and supporting details for each subtopic. List them in the box below the subtopic.

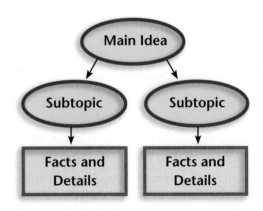

Practice

Read the passage below carefully. Then use the graphic organizer above to organize the information from the passage.

> Cultures change slowly over time. New ideas and new people can often lead to cultural change.
>
> Cultures often change as new ideas are introduced to a society. New ways of doing things, new inventions, and even new beliefs can all change a culture. One example of this is the spread of computer technology. As people adopted computers, they learned a new language and new work habits.
>
> Cultures also change when new people introduce their culture traits to a society. For example, as immigrants settle in the United States, they add new culture traits, like food, music, and clothing, to American culture.

Apply

Turn to Section 1 and read the passage titled Culture Regions. Draw a graphic organizer like the one above. Then follow the steps to organize the information you have read. The passage will have two or more subtopics. Add additional circles and rectangles for each additional subtopic you find.

Global Connections

If YOU lived there...

You live in Louisville, Kentucky, and you have never traveled out of the United States. However, when you got ready for school this morning, you put on a T-shirt made in Guatemala and jeans made in Malaysia. Your shoes came from China. You rode to school on a bus with parts manufactured in Mexico. At school, your class even took part in a discussion with students in Canada.

What makes your global connections possible?

> **BUILDING BACKGROUND** Trade and technology have turned the world into a "global village." People around the world wear clothes, eat foods, and use goods made in other countries. Global connections are bringing people around the world closer than ever before.

Globalization

In just seconds an e-mail message sent by a teenager in India beams all the way to a friend in London. A band in Seattle releases a new CD that becomes popular in China. People from New York to Singapore respond to a crisis in Brazil. These are all examples of **globalization**, the process in which countries are increasingly linked to each other through culture and trade.

What caused globalization? Improvements in transportation and communication over the past 100 years have brought the world closer together. Airplanes, telecommunications, and the Internet allow us to communicate and travel the world with ease. As a result, global culture and trade have increased.

Popular Culture

What might you have in common with a teenager in Japan? You probably have more in common than you think. You may use similar technology, wear similar clothes, and watch many of the same movies. You share the same global popular culture.

What You Will Learn...

Main Ideas

1. Globalization links the world's countries together through culture and trade.
2. The world community works together to solve global conflicts and crises.

The Big Idea

Fast, easy global connections have made cultural exchange, trade, and a cooperative world community possible.

Key Terms

globalization, *p. 97*
popular culture, *p. 98*
interdependence, *p. 99*
United Nations (UN), *p. 99*
humanitarian aid, *p. 100*

TAKING NOTES As you read, take notes on globalization and the world community. Use a graphic organizer like the one below to take notes.

Globalization	World Community

More and more, people around the world are linked through popular culture. **Popular culture** refers to culture traits that are well known and widely accepted. Food, sports, music, and movies are all examples of our popular culture.

The United States has great influence on global popular culture. For example, American soft drinks are sold in almost every country in the world. Many popular American television shows are broadcast internationally. English has become the major global language. One-quarter of the world's people speak English. It has become the main language for international music, business, science, and education.

At the same time, the United States is influenced by global culture. Martial arts movies from Asia attract large audiences in the United States. Radio stations in the United States play music by African, Latin American, and European musicians. We even adopt many foreign words, like *sushi* and *plaza*, into English.

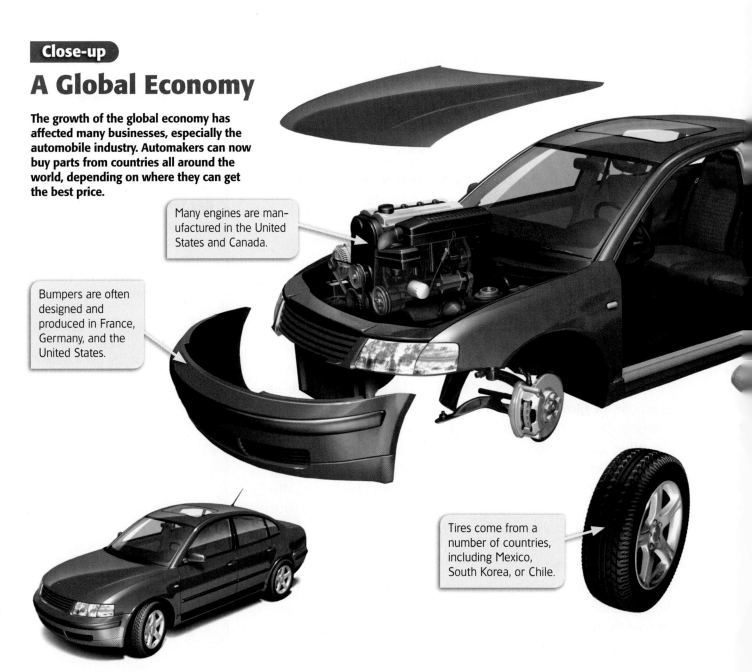

Close-up

A Global Economy

The growth of the global economy has affected many businesses, especially the automobile industry. Automakers can now buy parts from countries all around the world, depending on where they can get the best price.

Many engines are man-ufactured in the United States and Canada.

Bumpers are often designed and produced in France, Germany, and the United States.

Tires come from a number of countries, including Mexico, South Korea, or Chile.

Global Trade

Globalization not only links the world's people, but it also connects businesses and affects trade. For centuries, societies have traded with each other. Improvements in transportation and communication have made global trade quicker and easier. For example, a shoe retailer in Chicago can order the sneakers she needs on a Web site from a company in China. The order can be flown to Chicago the next day and sold to customers that afternoon.

The expansion of global trade has increased interdependence among the world's countries. **Interdependence** is a relationship between countries in which they rely on one another for resources, goods, or services. Many companies in one country often rely on goods and services produced in another country. For example, automakers in Europe might purchase auto parts made in the United States or Japan. Consumers also rely on goods produced elsewhere. For example, American shoppers buy bananas from Ecuador and tomatoes from Mexico. Global trade gives us access to goods from around the world.

READING CHECK Finding Main Ideas How has globalization affected the world?

Many cars feature windows manufactured in Venezuela or the United States.

Seats are sometimes assembled in Japan from covers sewn in Mexico.

ANALYSIS SKILL ANALYZING VISUALS

From what different countries do automotive parts often originate?

A World Community

Some people call our world a global village. What do you think this means? Because of globalization, the world seems smaller. Places are more connected. What happens in one part of the world can affect the entire planet. Because of this, the world community works together to promote cooperation among countries in times of conflict and crisis.

The world community encourages cooperation by working to resolve global conflicts. From time to time, conflicts erupt among the countries of the world. Wars, trade disputes, and political disagreements can threaten the peace. Countries often join together to settle such conflicts. In 1945, for example, 51 nations created the United Nations. The **United Nations (UN)** is an organization of the world's countries that promotes peace and security around the globe.

The world community also promotes cooperation in times of crisis. A disaster may leave thousands of people in need.

FOCUS ON READING

What is the main idea of this paragraph? What facts are used to support that idea?

HISTORIC DOCUMENT
The Charter of the United Nations

Created in 1945, the United Nations is an organization of the world's countries that works to solve global problems. The Charter of the United Nations outlines the goals of the UN, some of which are included here.

> We the Peoples of the United Nations Determined ...
>
> to save succeeding generations from the scourge [terror] of war ...
>
> to practice tolerance and live together in peace with one another as good neighbors, and
>
> to unite our strength to maintain international peace and security, and
>
> to ensure ... that armed forces shall not be used, save [except] in the common interest, and
>
> to employ international machinery [systems] for the promotion of the economic and social advancement of all peoples,
>
> Have Resolved to Combine our Efforts to Accomplish these Aims.
>
> —*from the Charter of the United Nations*

ANALYSIS SKILL **ANALYZING PRIMARY SOURCES**

What are some of the goals of the United Nations?

Earthquakes, floods, and drought can cause crises around the world. Groups from many nations often come together to provide **humanitarian aid**, or assistance to people in distress.

Organizations representing countries around the globe work to help in times of crisis. For example, in 2004 a tsunami, or huge tidal wave, devastated parts of Southeast Asia. Many organizations, like the United Nations Children's Fund (UNICEF) and the International Red Cross, stepped in to provide humanitarian aid to the victims of the tsunami. Some groups lend aid to refugees, or people who have been forced to flee their homes. Groups like Doctors Without Borders give medical aid to those in need around the world.

READING CHECK **Analyzing** How has globalization promoted cooperation?

SUMMARY In this section you learned how globalization links the countries of the world through shared culture and trade. Globalization allows organizations around the world to work together. They often solve conflicts and provide humanitarian aid.

Section 4 Assessment

go.hrw.com
Online Quiz
KEYWORD: SGA7 HP4

Reviewing Key Ideas, Terms, and Places
1. **a. Describe** What is **globalization**?
   b. Make Inferences How has **popular culture** influenced countries around the world?
   c. Evaluate In your opinion, has globalization hurt or helped the people of the world?
2. **a. Define** What is **humanitarian aid**?
   b. Draw Conclusions How has globalization promoted cooperation among countries?
   c. Predict What types of problems might lead to international cooperation?

Critical Thinking
3. **Identifying Cause and Effect** Use your notes and the graphic organizer at right to identify the effects that globalization has on our world.

Globalization → Effects / Effects / Effects

FOCUS ON WRITING
4. **Writing about Global Connections** What aspects of globalization might you include in your poster? Jot down your ideas in your notebook.

Geography's Impact
video series
Review the video to answer the closing question:
Why do you think some peoples must work to preserve their cultures in the modern world?

Visual Summary

Use the visual summary below to help you review the main ideas of the chapter.

QUICK FACTS

The world has many different cultures, or shared beliefs and practices.

The world's people practice different economic activities and systems.

Globalization brings people around the world closer than ever before.

Reviewing Vocabulary, Terms, and Places

Choose one word from each word pair to correctly complete each sentence below.

1. Members of a/an _____ often share the same religion, traditions, and language. **(ethnic group/population)**

2. People in a _____ are free to buy and sell goods as they please. **(command economy/market economy)**

3. Organizations like the International Red Cross provide _____ to people in need around the world. **(humanitarian aid/cultural diffusion)**

4. _____, the process of moving from one place to live in another, is a cause of population change. **(Population density/Migration)**

5. A country with a strong economy and a high standard of living is considered a _____. **(developed country/developing country)**

Comprehension and Critical Thinking

SECTION 1 *(Pages 80–85)*

6. **a. Describe** What is cultural diversity?

   b. Analyze What causes cultures to change over time?

   c. Elaborate Describe some of the culture traits practiced by people in your community.

SECTION 2 *(Pages 86–90)*

7. **a. Describe** What does population density tell us about a place?

   b. Draw Conclusions Why do certain areas attract large populations?

   c. Elaborate Why do you think it is important for geographers to study population trends?

SECTION 3 *(Pages 91–95)*

8. **a. Recall** What is a command economy?

SECTION 3 (continued)

b. Make Inferences Why might developing countries have only primary and secondary economic activities?

c. Evaluate Do you think government is important in our everyday lives? Why or why not?

SECTION 4 (Pages 97–100)

9. a. Describe How have connections among the world's countries improved?

b. Analyze What impact has globalization had on world trade and culture?

c. Evaluate What do you think has been the most important result of globalization? Why?

Social Studies Skills

10. Organizing Information Practice organizing information by creating a graphic organizer for Section 3. Use the main ideas on the first page of the section for your large circles. Then write the subtopics under each main idea. Finally, identify supporting details for each subtopic.

FOCUS ON READING AND WRITING

Understanding Main Ideas *Read the paragraph below carefully, then write out the main idea of the paragraph.*

11. The ancient Greeks were the first to practice democracy. Since then many countries have adopted democratic government. The United Kingdom, South Korea, and Ghana all practice democracy. Democracy is the most widely used government in the world today.

Creating a Poster *Use your notes and the instructions below to help you create a poster.*

12. Review your notes about the world's cultures, populations, governments, and economies. Then select a subject for your poster. On a large sheet of paper, write a title that identifies your topic. Decorate your poster with illustrations that relate to your main idea. Write a short caption explaining each image. Be sure to use words and images that will grab your audience's attention and clearly express your main idea.

Using the Internet

go.hrw.com
KEYWORD: SGA7 CH4

13. Activity: Writing a Report Population changes have a huge impact on the world around us. Countries around the world must deal with shrinking populations, growing populations, and other population issues. Enter the activity keyword and explore the issues surrounding global population. Then imagine you have been asked to report on global population trends to the United Nations. Write a report in which you identify world population trends and their impact on the world today.

Map Activity ★Interactive

Population Density *Use the map below to answer the questions that follow.*

14. What letter on the map indicates the least crowded area?

15. What letter on the map indicates the most densely crowded area?

16. Which letter indicates a region with 260–520 people per square mile (100–200 people per square km)?

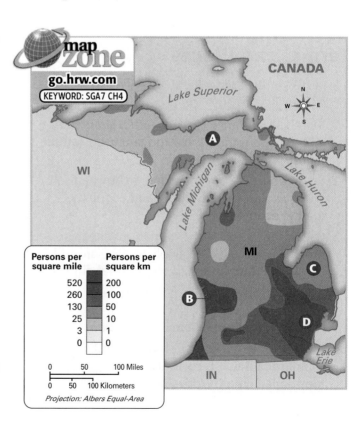

DIRECTIONS: Read questions 1 through 7 and write the letter of the best response. Then read question 8 and write your own well-constructed response.

1 **Which of the following is *most likely* a culture trait?**

A religion

B population density

C interdependence

D cultural diffusion

2 **What developments led to the rapid increase in world population in the last 200 years?**

A a decline in migration

B improvements in technology and communication

C a decrease in standard of living

D improvements in health care and agriculture

3 **Which economic system is used in the United States?**

A market economy

B command economy

C traditional economy

D domestic economy

4 **A government in which a single, powerful ruler exerts complete control is a**

A Communist state.

B democracy.

C dictatorship.

D republic.

5 **Global connections have improved as a result of**

A population growth.

B cultural diversity.

C the spread of democratic government.

D improvements in technology.

Developed and Developing Countries

Country	Per Capita GDP (U.S. $)	Life Expectancy at Birth	TVs per 1,000 People
Cameroon	$1,900	50.9	34
Singapore	$27,800	81.6	341
Ukraine	$6,300	69.7	433
Uruguay	$14,500	76.1	531

6 **Which of the countries in the chart above is *most likely* a developed country?**

A Cameroon

B Singapore

C Ukraine

D Uruguay

7 **Which of the following is an example of economic interdependence?**

A Cattle ranchers in Oklahoma sell beef to grocery stores in Maryland.

B Students in Germany use the Internet to communicate with scientists in Brazil.

C Construction companies in Canada build skyscrapers with steel imported from the United States.

D Immigrants from Russia settle in London.

8 **Extended Response** Using the data in the chart above, write a paragraph in which you compare and contrast the standard of living in Ukraine and Singapore.

Explaining a Process

How does soil renewal work? How do cultures change? Often the first question we ask about something is how it works or what process it follows. One way we can answer these questions is by writing an explanation.

Assignment
Write a paper explaining one of these topics:
- how water recycles on Earth
- how agriculture developed

1. Prewrite

Choose a Process
- Choose one of the topics above to write about.
- Turn your topic into a big idea, or thesis. For example, your big idea might be "Water continually circulates from Earth's surface to the atmosphere and back."

> **TIP** **Organizing Information** Explanations should be in a logical order. You should arrange the steps in the process in chronological order, the order in which the steps take place.

Gather and Organize Information
- Look for information about your topic in your textbook, in the library, or on the Internet.
- Start a plan to organize support for your big idea. For example, look for the individual steps of the water cycle.

2. Write

Use a Writer's Framework

> **A Writer's Framework**
>
> **Introduction**
> - Start with an interesting fact or question.
> - Identify your big idea.
>
> **Body**
> - Create at least one paragraph for each point supporting the big idea. Add facts and details to explain each point.
> - Use chronological order or order of importance.
>
> **Conclusion**
> - Summarize your main points in your final paragraph.

3. Evaluate and Revise

Review and Improve Your Paper
- Re-read your paper and make sure you have followed the framework.
- Make the changes needed to improve your paper.

Evaluation Questions for an Explanation of a Process
1. Do you begin with an interesting fact or question?
2. Does your introduction identify your big idea? Does it provide any background information your readers might need?
3. Do you have at least one paragraph for each point you are using to support the big idea?
4. Do you include facts and details to explain and illustrate each point?
5. Do you use chronological order or order of importance to organize your main points?

4. Proofread and Publish

Give Your Explanation the Finishing Touch
- Make sure you have capitalized the first word in every sentence.
- Check for punctuation at the end of every sentence.
- Think of a way to share your explanation.

5. Practice and Apply

Use the steps and strategies outlined in this workshop to write your explanation. Share your paper with others and find out whether the explanation makes sense to them.

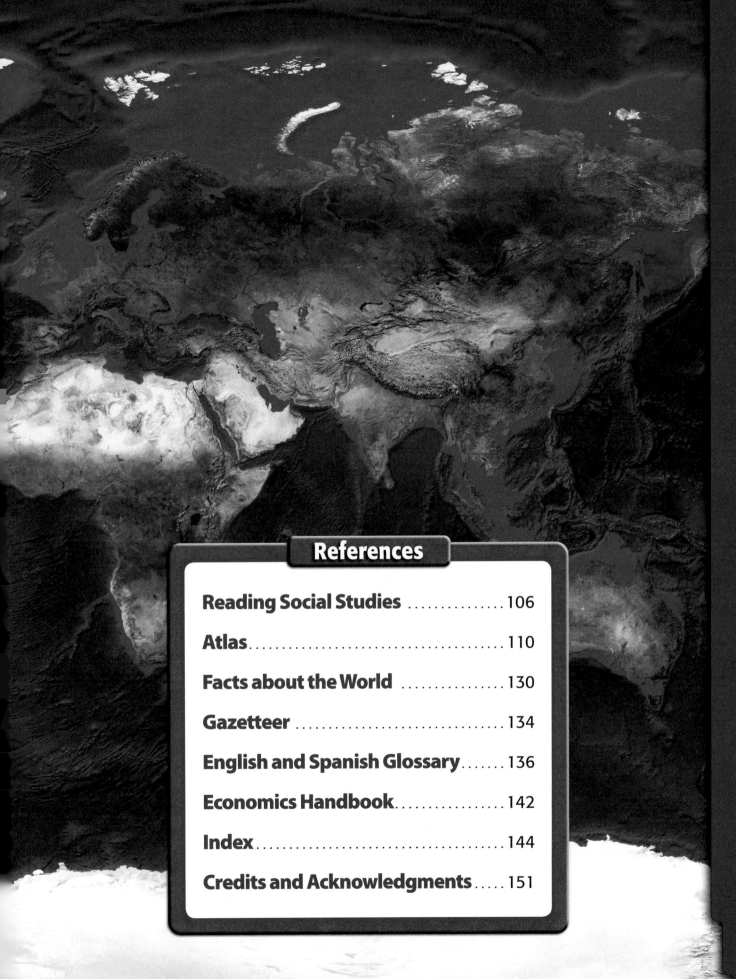

References

Reading Social Studies 106

Atlas 110

Facts about the World 130

Gazetteer 134

English and Spanish Glossary 136

Economics Handbook 142

Index 144

Credits and Acknowledgments 151

Using Prior Knowledge

FOCUS ON READING

When you put together a puzzle, you search for pieces that are missing to complete the picture. As you read, you do the same thing when you use prior knowledge. You take what you already know about a subject and then add the information you are reading to create a full picture. The example below shows how using prior knowledge about computer mapping helped one reader fill in the pieces about how geographers use computer mapping.

In the past, maps were always drawn by hand. Many were not very accurate. Today, though, most maps are made using computers and satellite images. Through advances in mapmaking, we can make accurate maps on almost any scale, from the whole world to a single neighborhood, and keep them up to date.

From Section 3, The Branches of Geography

Computer Mapping	
What I know before reading	What else I learned
• My dad uses the computer to get a map for trips. • I can find maps on the Internet of states and countries.	• Maps have not always been very accurate. • Computers help make new kinds of maps that are more than just cities and roads. • These computer maps are an important part of geography.

YOU TRY IT!

Draw a chart like the one above. Think about what you know about satellite images and list this prior knowledge in the left column of your chart. Then read the passage below. Once you have read it, add what you learned about satellite images to the right column.

Much of the information gathered by these satellites is in the form of images. Geographers can study these images to see what an area looks like from above Earth. Satellites also collect information that we cannot see from the planet's surface. The information gathered by satellites helps geographers make accurate maps.

From Section 1, Studying Geography

Using Word Parts

FOCUS ON READING

Many English words are made up of several word parts: roots, prefixes, and suffixes. A root is the base of the word and carries the main meaning. A prefix is a letter or syllable added to the beginning of a root. A suffix is a letter or syllable added to the end to create new words. When you come across a new word, you can sometimes figure out the meaning by looking at its parts. Below are some common word parts and their meanings.

Common Roots		
Word Root	**Meaning**	**Sample Words**
-graph-	write, writing	autograph, biography
-vid-, -vis-	see	videotape, visible

Common Prefixes		
Prefix	**Meaning**	**Sample Words**
geo-	earth	geology
inter-	between, among	interpersonal, intercom
in-	not	ineffective
re-	again	restate, rebuild

Common Suffixes		
Suffix	**Meaning**	**Sample Words**
-ible	capable of	visible, responsible
-less	without	penniless, hopeless
-ment	result, action	commitment
-al	relating to	directional
-tion	the act or condition of	rotation, selection

YOU TRY IT!

Read the following words. First separate any prefixes or suffixes and identify the word's root. Use the chart above to define the root, the prefix, or the suffix. Then write a definition for each word.

geography	**visualize**	**movement**
seasonal	**reshaping**	**interact**
regardless	**separation**	**invisible**

Understanding Cause and Effect

FOCUS ON READING

Learning to identify causes and effects can help you understand geography. A **cause** is something that makes another thing happen. An **effect** is the result of something else that happened. A cause may have several effects, and an effect may have several causes. In addition, as you can see in the example below, causes and effects may occur in a chain. Then, each effect in turn becomes the cause for another event.

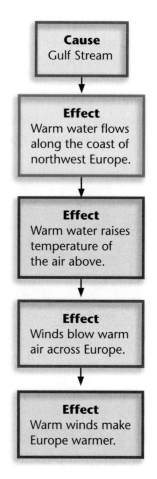

Cause
Gulf Stream

Effect
Warm water flows along the coast of northwest Europe.

Effect
Warm water raises temperature of the air above.

Effect
Winds blow warm air across Europe.

Effect
Warm winds make Europe warmer.

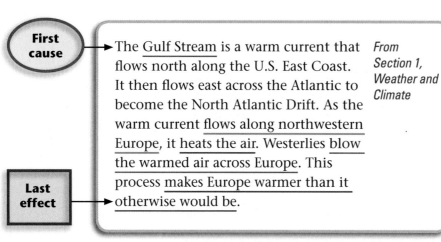

First cause → The Gulf Stream is a warm current that flows north along the U.S. East Coast. It then flows east across the Atlantic to become the North Atlantic Drift. As the warm current flows along northwestern Europe, it heats the air. Westerlies blow the warmed air across Europe. This process makes Europe warmer than it otherwise would be. ← **Last effect**

From Section 1, Weather and Climate

YOU TRY IT!

Read the following sentences, and then use a graphic organizer like the one below right to analyze the cause and effects. Create as many boxes as you need to list the causes and effects.

Mountains also create wet and dry areas. . . A mountain forces wind blowing against it to rise. As it rises, the air cools and precipitation falls as rain or snow. Thus, the side of the mountain facing the wind is often green and lush. However, little moisture remains for the other side. This effect creates a rain shadow.

From Section 1, Weather and Climate

Cause → **Effect** → **Effect** → **Effect**

Chapter 4 The World's People

Understanding Main Ideas

FOCUS ON READING

Main ideas are like the hub of a wheel. The hub holds the wheel together, and everything circles around it. In a paragraph, the main idea holds the paragraph together and all the facts and details revolve around it. The main idea is usually stated clearly in a topic sentence, which may come at the beginning or end of a paragraph. Topic sentences always summarize the most important idea of a paragraph.

To find the main idea, ask yourself what one point is holding the paragraph together. See how the main idea in the following example holds all the details from the paragraph together.

A single country may also include more than one culture region within its borders. Mexico is one of many countries that is made up of different culture regions. People in northern Mexico and southern Mexico, for example, have different culture traits. The culture of northern Mexico tends to be more modern, while traditional culture remains strong in southern Mexico.

From Section 1, Culture

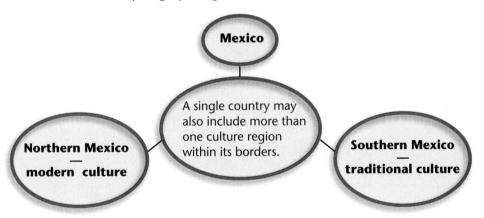

YOU TRY IT!

Read the following paragraph, and then use a graphic organizer like the one above to identify the main idea. Create as many circles as you need to list the supporting facts and details.

At the same time, the United States is influenced by global culture. Martial arts movies from Asia attract large audiences in the United States. Radio stations in the United States play music by African, Latin American, and European musicians. We even adopt many foreign words, like *sushi* and *plaza*, into English.

From Section 4, Global Connections

READING SOCIAL STUDIES

United States: Physical

Strait of Juan de Fuca

Puget Sound

Mount Rainier
14,410 ft
(4,392 m)

Franklin D. Roosevelt Lake

Columbia River

C O A S T   R A N G E S

C A S C A D E   R A N G E

Columbia Plateau

Bitterroot Range

Salmon River Mts.

Sawtooth Mts.

CONTINENTAL

Snake River

Klamath River

Goose Lake

Shasta Lake

Pyramid Lake

Lake Tahoe

GREAT

BASIN

Great Salt Lake

Utah Lake

Wasatch Range

Grand Tetons

Gannett Peak
13,804 ft
(4,207 m)

Wind River Range

Uinta Mts.

Green River

R O C K Y

Lewis Range

Milk River

Missouri River

Fort Peck Lake

Yellowstone River

Bighorn Mts.

Bighorn River

Powder River

Yellowstone River

DIVIDE

M O U N T A I N S

Front Range

Lake Sakakawea

Lake Oahe

Black Hills

Cheyenne River

White River

Niobrara River

North Platte River

South Platte River

Platte River

Republican River

Smoky Hill River

G R E A T

I N T E R

P L A I N S

Lake Oahe

James River

SIERRA NEVADA

Central Valley

Sacramento River

San Joaquin River

Cape Mendocino

San Francisco Bay

Monterey Bay

Coast Ranges

Mount Whitney
14,494 ft
(4,419 m)

Death Valley

Mojave Desert

COLORADO

PLATEAU

Mount Elbert
14,433 ft
(4,400 m)

Pikes Peak
14,110 ft
(4,301 m)

San Luis Valley

Sangre De Cristo Mts.

Painted Desert

Lake Powell

Colorado River

Grand Canyon

Lake Mead

San Juan River

DIVIDE

Gila River

Salton Sea

Imperial Valley

Channel Islands

PACIFIC

OCEAN

Sonoran Desert

CONTINENTAL

Rio Grande

Gulf of California

Pecos River

Amistad Reservoir

Canadian River

Colorado River

Nueces River

MEXICO

To understand the relative locations of Alaska and Hawaii, as well as the vast distances separating them from the rest of the United States, see the world map.

Kauai

Niihau

Oahu

Molokai

HAWAII

PACIFIC
OCEAN

Lanai

Maui

Kahoolawe

Mauna Kea
13,796 ft
(4,206 m)

Hawaii

0   75   150 Miles

0   75   150 Kilometers

Projection: Mercator

ARCTIC OCEAN

RUSSIA

Arctic Circle

Bering Strait

BROOKS RANGE

Yukon River

Tanana River

St. Lawrence Island

St. Matthew Island

Nunivak Island

Kuskokwim River

ALASKA RANGE

Mount McKinley
20,320 ft
(6,194 m)

CANADA

Bering Sea

Attu Island

ALEUTIAN ISLANDS

PACIFIC
OCEAN

Gulf of Alaska

Kodiak Island

Alexander Archipelago

0   250   500 Miles

0   250   500 Kilometers

Projection: Albers Equal Area

CANADA

Red River

Isle Royale

Mesabi Range

Lake Superior

Minnesota River

Mississippi River

Wisconsin River

Lake Michigan

Lake Huron

Des Moines River

Lake Ontario

St. Lawrence River

St. Lawrence Seaway

Lake Champlain

Adirondack Mts.

Green Mts.

White Mts.

Long Island Sound

Long Island

Cape Cod

Penobscot River

St. John River

Long Fellow Mts.

Connecticut River

Hudson River

ALLEGHENY PLATEAU

Catskill Mts.

Susquehanna River

Allegheny R.

Delaware River

Delaware Bay

40°N

70°W

Kansas R.

P L A I N S

Illinois River

Wabash River

Scioto River

Ohio River

Monongahela R.

Kanawha River

Potomac River

James River

APPALACHIAN MOUNTAINS

Chesapeake Bay

ATLANTIC OCEAN

Lake of the Ozarks

OZARK PLATEAU

Lake Barkley

Cumberland River

Cumberland Plateau

Great Smoky Mts.

BLUE RIDGE MOUNTAINS

Roanoke River

Pamlico Sound

Cape Hatteras

35°N

Keystone Lake

Arkansas River

White River

Kentucky Lake

Tennessee River

P I E D M O N T

Ouachita Mts.

Lake Texoma

Eufaula Lake

Tombigbee River

Coosa River

Oconee River

Savannah River

Trinity River

Saline River

Red River

Mississippi River

Pearl River

Alabama R.

Chattahoochee River

Altamaha River

Sea Islands

P L A I N

Brazos River

Toledo Bend Reservoir

G U L F   C O A S T A L

Chandeleur Islands

Mississippi Delta

Okefenokee Swamp

Cape Canaveral

FLORIDA PENINSULA

N
W   E
S

Gulf of Mexico

Lake Okeechobee

BAHAMAS

25°N

The Everglades

Cape Sable

Florida Keys

Straits of Florida

85°W

80°W

75°W

90°W

95°W

ELEVATION

Feet	Meters
13,120	4,000
6,560	2,000
1,640	500
656	200
(Sea level) 0	0 (Sea level)
Below sea level	Below sea level

0   100   200 Miles

0   100   200 Kilometers

Projection: Albers Equal Area

ATLAS

United States: Political

Strait of Juan de Fuca

Seattle
Tacoma
Olympia ★
WASHINGTON
Spokane
Franklin D. Roosevelt Lake
Pend Oreille
Flathead Lake

45°N

Portland
Columbia River
Salem ★
Eugene
OREGON

Great Falls
Helena ★
MONTANA
Fort Peck Lake
Missouri River
Yellowstone River
Billings

NORTH DAKOTA
Lake Sakakawea
Bismarck ★

Boise ★
Sun Valley
IDAHO
Snake River
Pocatello

Yellowstone Lake
WYOMING
Cheyenne ★

Lake Oahe
SOUTH DAKOTA
Pierre ★
Rapid City

40°N

Cape Mendocino
Goose Lake
Shasta Lake
Sacramento River
Pyramid Lake
NEVADA
Reno
Carson City ★
Lake Tahoe

Great Salt Lake
Ogden
Salt Lake City ★
Provo
Utah Lake
UTAH
Green River

Boulder
Vail
Denver ★
Aspen
Colorado Springs
COLORADO
Pueblo

NEBRASKA
Platte River

125°W

Berkeley
Oakland
San Francisco
San Francisco Bay
San Jose
Monterey Bay
Sacramento
San Joaquin River

35°N

Fresno
CALIFORNIA

Las Vegas

Lake Powell
Lake Mead
Colorado River

Arkansas River
KANSAS

Santa Barbara
Ventura
Los Angeles
Long Beach
Anaheim
Santa Ana
San Diego
Channel Islands
Riverside
Palm Springs
Salton Sea

Flagstaff
ARIZONA
Phoenix ★
Gila River
Casa Grande
Tucson

Taos
Santa Fe ★
Albuquerque
NEW MEXICO
Las Cruces
El Paso

Canadian River
OKLAHOMA
Oklahoma City ★
Lawto...
Amarillo

PACIFIC OCEAN

30°N

120°W

Gulf of California

To understand the relative locations of Alaska and Hawaii, as well as the vast distances separating them from the rest of the United States, see the world map.

Lubbock
Brazos River
Abilene
Fort Wor...
Midland
Odessa
TEXAS
Pecos River
Colorado River

Amistad Reservoir
Austin

Rio Grande
San Antonio

MEXICO

Corpus Christi
Laredo
Pad... Isla...

HAWAII Inset

Kauai
Niihau
Oahu
Molokai
HAWAII
Honolulu ★
PACIFIC OCEAN
Lanai
Maui
Kahoolawe
Hilo
Hawaii
22°N
155°W
19°N

```
0      75      150 Miles
0      75      150 Kilometers
Projection: Mercator
```

ALASKA Inset

ARCTIC OCEAN
Arctic Circle
RUSSIA
Bering Strait
Nome
Yukon River
Fairbanks
CANADA
St. Lawrence Island
St. Matthew Island
Nunivak Island
ALASKA
Anchorage
Valdez
Skagway
Juneau ★
Kodiak Island
Gulf of Alaska
Alexander Archipelago
Bering Sea
Attu Island
PACIFIC OCEAN
ALEUTIAN ISLANDS

160°W
170°W
180°
170°E
55°N
50°N

```
0      250      500 Miles
0      250      500 Kilometers
Projection: Albers Equal Area
```

CANADA

ATLAS

MINNESOTA
Grand Forks
Fargo
Duluth
Superior
Lake Superior
Marquette
Sault Ste. Marie
WISCONSIN
MICHIGAN
Minneapolis
St. Paul
Green Bay
Lake Michigan
Lake Huron
Madison
Milwaukee
Grand Rapids
Saginaw
Lansing
Detroit
Ann Arbor
Sioux Falls
Sioux City
IOWA
Cedar Rapids
Davenport
Des Moines
Rockford
Chicago
Gary
South Bend
Fort Wayne
Toledo
Cleveland
Youngstown
Akron
OHIO
Columbus
Dayton
Cincinnati
maha
ncoln
Peoria
INDIANA
Springfield
Indianapolis
MISSOURI
Kansas City
Kansas City
ILLINOIS
St. Louis
East St. Louis
Louisville
Evansville
Frankfort
Lexington
WEST VIRGINIA
Charleston
VIRGINIA
Richmond
Newport News
Norfolk
Virginia Beach
opeka
Jefferson City
ichita
Lake of the Ozarks
Springfield
KENTUCKY
Lake Barkley
Keystone Lake
Tulsa
Fayetteville
Nashville
Knoxville
Asheville
Charlotte
Greensboro
Durham
Raleigh
Cape Hatteras
NORTH CAROLINA
Winston-Salem
ufaula Lake
ARKANSAS
Little Rock
Pine Bluff
Memphis
Chattanooga
TENNESSEE
Greenville
SOUTH CAROLINA
Columbia
Charleston
Lake Texoma
Huntsville
Atlanta
Birmingham
Macon
Savannah
allas
MISSISSIPPI
Vicksburg
Jackson
Meridian
Montgomery
ALABAMA
GEORGIA
Columbus
aco
Toledo Bend Reservoir
LOUISIANA
Shreveport
Baton Rouge
Biloxi
Mobile
Pensacola
Tallahassee
Jacksonville
Beaumont
Houston
New Orleans
Chandeleur Islands
Gainesville
FLORIDA
Galveston
Orlando
Tampa
St. Petersburg
Lake Okeechobee
Gulf of Mexico
Fort Myers
Fort Lauderdale
Miami
Cape Sable
Florida Keys
Straits of Florida
BAHAMAS

MAINE
Augusta
Lake Champlain
Burlington
Montpelier
Portland
VT
NH
Concord
Manchester
Hudson
Lake Ontario
Rochester
Syracuse
Albany
Springfield
MA
Boston
Worcester
Providence
Cape Cod
Buffalo
NEW YORK
Hartford
CT
RI
New Haven
Long Island Sound
Bridgeport
Jersey City
Yonkers
Long Island
Lake Erie
Erie
Susquehanna River
PENNSYLVANIA
Allentown
Newark
New York City
Trenton
Harrisburg
Philadelphia
Camden
NJ
Atlantic City
Pittsburgh
Baltimore
DE
Dover
MD
Washington, D.C.
Annapolis
Delaware Bay
Chesapeake Bay
ATLANTIC OCEAN

St. Lawrence River
Red River
Minnesota River
Mississippi River
Missouri River
Illinois River
Ohio River
Kentucky Lake
Savannah River
Chattahoochee River
Red River
Sea Islands
Cape Canaveral

Red River

N
W E
S

Cape Sable

40°N
35°N
30°N
25°N
70°W
75°W
80°W
85°W
90°W
95°W

National capital
State capitals
Other cities
0 100 200 Miles
0 100 200 Kilometers
Projection: Albers Equal Area

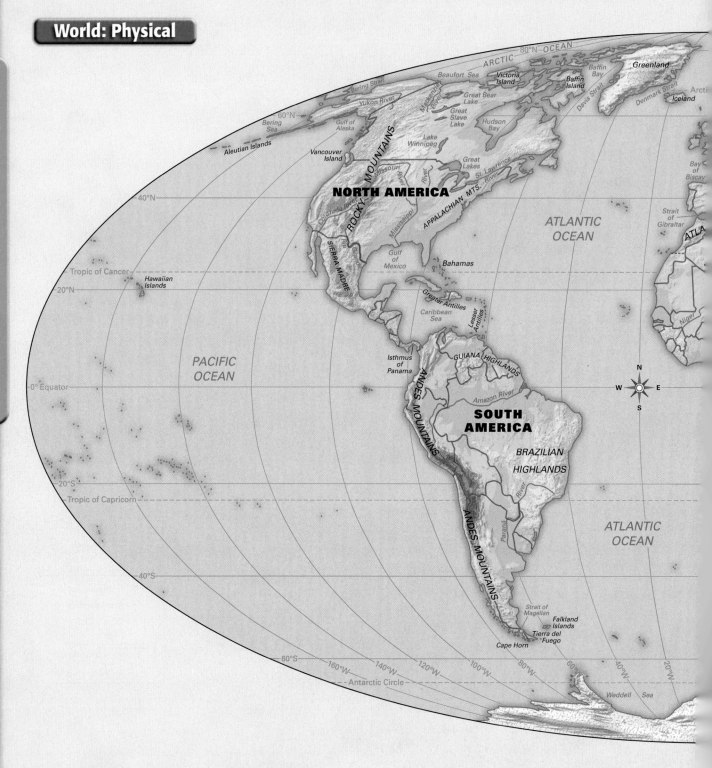

World: Physical

ATLAS

ARCTIC OCEAN
80°N

Beaufort Sea
Victoria Island
Baffin Bay
Baffin Island
Davis Strait
Greenland
Denmark Strait
Iceland
Arcti

Bering Strait
Yukon River
Mackenzie River
Great Bear Lake
Great Slave Lake
Hudson Bay

60°N
Bering Sea
Gulf of Alaska
ROCKY MOUNTAINS
Lake Winnipeg

Aleutian Islands
Vancouver Island
Missouri River
Great Lakes
St. Lawrence River

NORTH AMERICA
40°N
Colorado River
ROCKY MTS.
Mississippi
APPALACHIAN MTS.

Bay of Biscay
ATLA

SIERRA MADRE
Rio Grande
Gulf of Mexico
Bahamas

ATLANTIC OCEAN

Strait of Gibraltar
ATLA

Tropic of Cancer
Hawaiian Islands
20°N
Greater Antilles
Caribbean Sea
Lesser Antilles

Niger

PACIFIC OCEAN
Isthmus of Panama
GUIANA HIGHLANDS

0° Equator
ANDES MOUNTAINS
Amazon River

SOUTH AMERICA

N
W **E**
S

BRAZILIAN HIGHLANDS

20°S
Paraná River

Tropic of Capricorn

ATLANTIC OCEAN

40°S
ANDES MOUNTAINS

Strait of Magellan
Falkland Islands
Tierra del Fuego
Cape Horn

60°S
160°W 140°W 120°W 100°W 80°W 60°W 40°W 20°W
Antarctic Circle

Weddell Sea

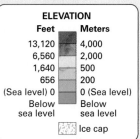

ELEVATION

Feet		Meters
13,120		4,000
6,560		2,000
1,640		500
656		200
(Sea level) 0		0 (Sea level)
Below sea level		Below sea level

Ice cap

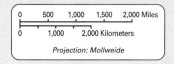

0 500 1,000 1,500 2,000 Miles
0 1,000 2,000 Kilometers

Projection: Mollweide

114 ATLAS

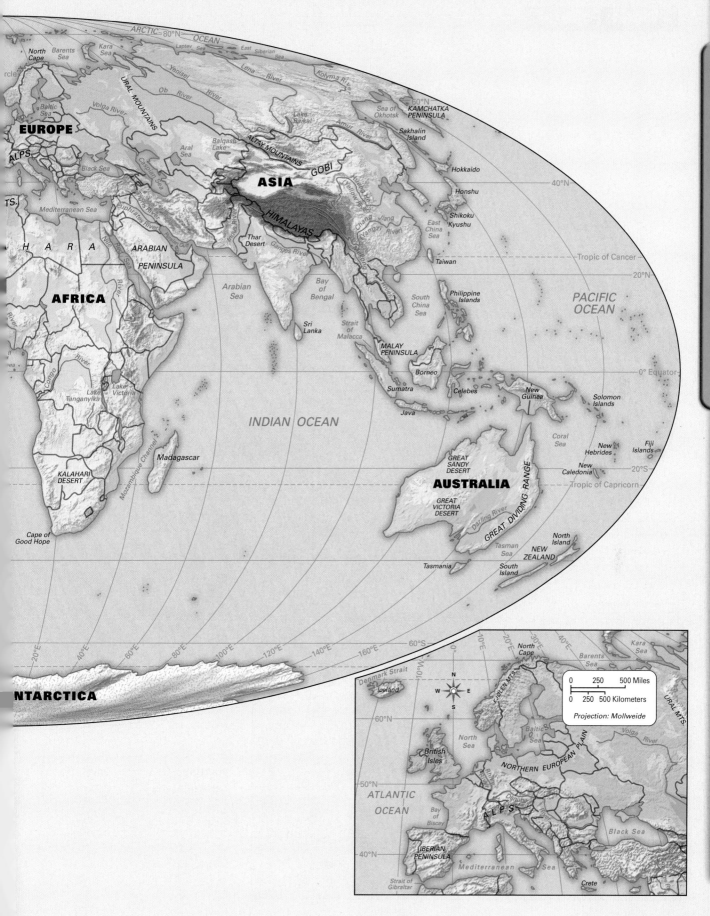

North Cape
Barents Sea
Kara Sea
ARCTIC 80°N OCEAN
Laptev Sea
East Siberian Sea
cle
Baltic Sea
URAL MOUNTAINS
Ob River
Yenisei River
Lena River
Kolyma R.
EUROPE
Black Sea
Volga River
Aral Sea
Balqash Lake
Caspian Sea
ALTAY MOUNTAINS
Lake Baikal
60°N
Sea of Okhotsk
KAMCHATKA PENINSULA
TS.
Mediterranean Sea
Euphrates River
GOBI
Amur River
Sakhalin Island
SAHARA
Tigris River
ASIA
Yellow River
Hokkaido
ARABIAN
Indus River
HIMALAYAS
Chang Jiang
Yangtze River
40°N
Honshu
Shikoku
Kyushu
PENINSULA
Gulf
Thar Desert
Ganges River
East China Sea
Nile River
Arabian Sea
Bay of Bengal
Taiwan
Tropic of Cancer
AFRICA
Mekong River
20°N
Sri Lanka
Strait of Malacca
South China Sea
Philippine Islands
PACIFIC OCEAN
Congo River
MALAY PENINSULA
Lake Victoria
Lake Tanganyika
Borneo
Sumatra
Celebes
0° Equator
Java
New Guinea
Solomon Islands
Mozambique Channel
INDIAN OCEAN
Madagascar
Coral Sea
New Hebrides
Fiji Islands
GREAT SANDY DESERT
GREAT DIVIDING RANGE
New Caledonia
20°S
KALAHARI DESERT
AUSTRALIA
Darling River
Tropic of Capricorn
GREAT VICTORIA DESERT
Cape of Good Hope
North Island
Tasman Sea
NEW ZEALAND
Tasmania
South Island
20°E
40°E
60°E
80°E
100°E
120°E
140°E
160°E
60°S
NTARCTICA

North Cape
20°E
20°E
40°E
Kara Sea
Denmark Strait
Barents Sea
N
Iceland
KJÖLEN MTS.
W E
S
URAL MTS.
0
250
500 Miles
0
250
500 Kilometers
Projection: Mollweide
60°N
North Sea
Baltic Sea
Volga River
British Isles
NORTHERN EUROPEAN PLAIN
ATLANTIC OCEAN
50°N
ALPS
Black Sea
Bay of Biscay
40°N
IBERIAN PENINSULA
Mediterranean Sea
Strait of Gibraltar
Crete

World: Political

ATLAS

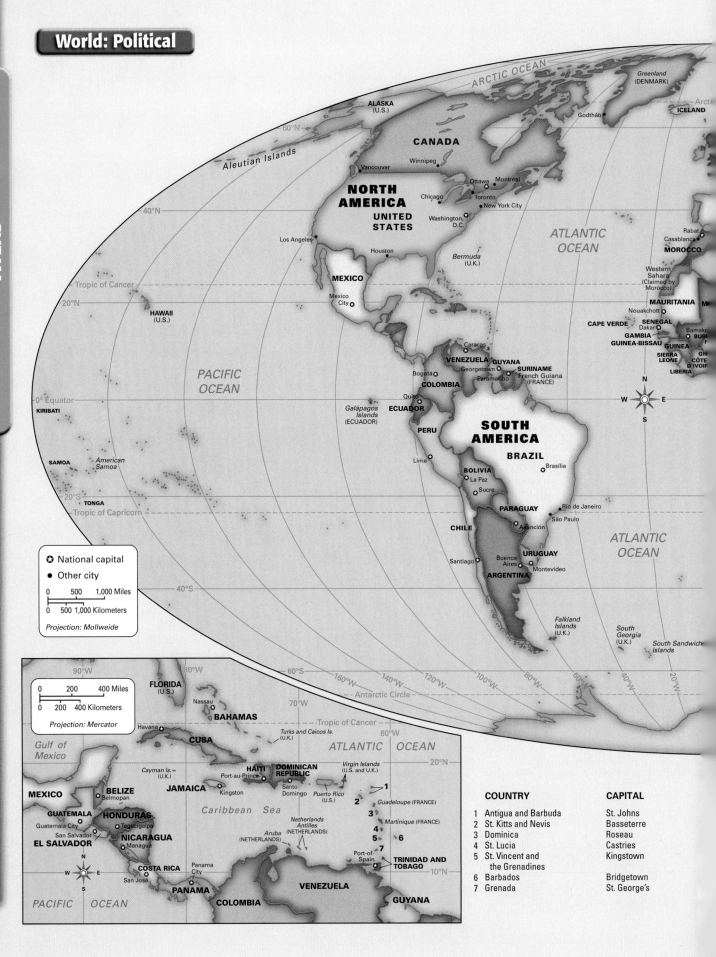

ARCTIC OCEAN

Greenland (DENMARK)

ICELAND

ALASKA (U.S.)

60°N

CANADA

Aleutian Islands

Vancouver
Winnipeg
Ottawa   Montreal
Chicago   Toronto

NORTH AMERICA

40°N

Washington, D.C.
New York City

ATLANTIC OCEAN

Rabat
Casablanca
MOROCCO

UNITED STATES

Los Angeles

Houston

Bermuda (U.K.)

Western Sahara (Claimed by Morocco)

Tropic of Cancer

MEXICO

MAURITANIA

20°N

Mexico City

Nouakchott

HAWAII (U.S.)

CAPE VERDE
Dakar
SENEGAL
GAMBIA
GUINEA-BISSAU
Bamako
BUR
GUINEA
SIERRA LEONE
GH
CÔTE D'IVOIF
LIBERIA

Caracas

VENEZUELA   GUYANA
Georgetown   SURINAME
Bogotá   Paramaribo
French Guiana (FRANCE)

KIRIBATI

PACIFIC OCEAN

0° Equator

COLOMBIA

Quito

N

W   E

S

Galápagos Islands (ECUADOR)

ECUADOR

PERU

SOUTH AMERICA

SAMOA

American Samoa

Lima

BRAZIL

Brasília

BOLIVIA
La Paz
Sucre

20°S

TONGA

PARAGUAY

Rio de Janeiro
São Paulo

Tropic of Capricorn

CHILE

Asunción

ATLANTIC OCEAN

URUGUAY

Santiago

Buenos Aires
Montevideo

ARGENTINA

○ National capital

● Other city

0   500   1,000 Miles

0   500 1,000 Kilometers

Projection: Mollweide

40°S

Falkland Islands (U.K.)

South Georgia (U.K.)

South Sandwich Islands

90°W   80°W   FLORIDA (U.S.)   60°W   160°W   140°W   120°W   100°W   80°W   60°W   40°W   20°W

Antarctic Circle

0   200   400 Miles

0   200   400 Kilometers

Projection: Mercator

Nassau

70°W

Tropic of Cancer

BAHAMAS

Havana

60°W

ATLANTIC OCEAN

Gulf of Mexico

CUBA

Turks and Caicos Is. (U.K.)

20°N

Cayman Is. (U.K.)

HAITI   DOMINICAN REPUBLIC

Virgin Islands (U.S. and U.K.)

MEXICO

BELIZE
Belmopan

JAMAICA
Kingston

Port-au-Prince
Santo Domingo

Puerto Rico (U.S.)

1

COUNTRY	CAPITAL
1 Antigua and Barbuda	St. Johns
2 St. Kitts and Nevis	Basseterre
3 Dominica	Roseau
4 St. Lucia	Castries
5 St. Vincent and the Grenadines	Kingstown
6 Barbados	Bridgetown
7 Grenada	St. George's

GUATEMALA
Guatemala City
San Salvador
EL SALVADOR

HONDURAS
Tegucigalpa

Caribbean Sea

Guadeloupe (FRANCE)

2

Martinique (FRANCE)

3

NICARAGUA
Managua

Netherlands Antilles (NETHERLANDS)

Aruba (NETHERLANDS)

4

5

6

N

W   E

S

COSTA RICA
San José

Panama City

7

Port-of-Spain

TRINIDAD AND TOBAGO

10°N

PACIFIC OCEAN

PANAMA

COLOMBIA

VENEZUELA

GUYANA

116 ATLAS

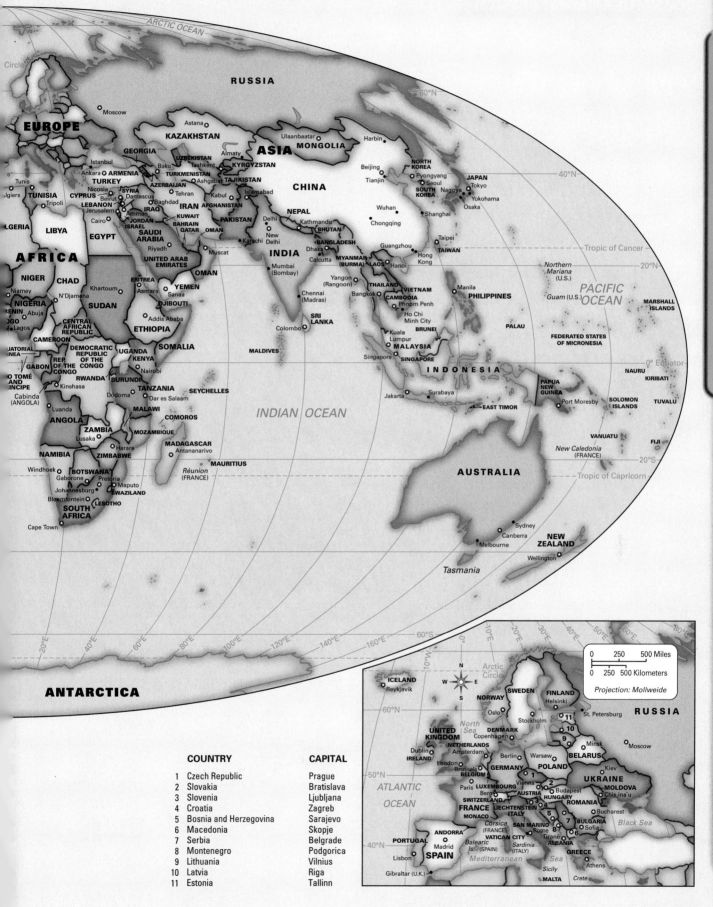

ARCTIC OCEAN

Circle

EUROPE

RUSSIA

Moscow

Astana

KAZAKHSTAN

Ulaanbaatar

MONGOLIA

Harbin

60°N

GEORGIA

ASIA

Almaty

UZBEKISTAN

Istanbul

Tashkent

KYRGYZSTAN

Beijing

NORTH
KOREA

40°N

Ankara

ARMENIA

TURKMENISTAN

Baku

TAJIKISTAN

Tianjin

Pyongyang

JAPAN

Tunis

TURKEY

AZERBAIJAN

Ashgabat

CHINA

SOUTH
KOREA

Seoul

Nagoya

Tokyo

Nicosia

Kabul

Wuhan

Osaka

Yokohama

Igiers

CYPRUS

SYRIA

Tehran

AFGHANISTAN

NEPAL

Shanghai

TUNISIA

LEBANON

Beirut

Damascus

IRAN

Islamabad

Chongqing

Tripoli

Jerusalem

Baghdad

Kathmandu

Amman

IRAQ

KUWAIT

PAKISTAN

Delhi

BHUTAN

Taipei

Cairo

JORDAN

BAHRAIN

New
Delhi

Dhaka

TAIWAN

Tropic of Cancer

LGERIA

ISRAEL

SAUDI

QATAR

OMAN

BANGLADESH

Guangzhou

20°N

LIBYA

EGYPT

ARABIA

Karachi

INDIA

Calcutta

MYANMAR
(BURMA)

Hong
Kong

Riyadh

UNITED ARAB

Muscat

Mumbai
(Bombay)

LAOS

Hanoi

AFRICA

EMIRATES

OMAN

Yangon
(Rangoon)

VIETNAM

Chennai
(Madras)

THAILAND

Manila

Northern
Mariana
(U.S.)

PACIFIC
OCEAN

NIGER

CHAD

ERITREA

YEMEN

Bangkok

CAMBODIA

PHILIPPINES

Guam (U.S.)

MARSHALL
ISLANDS

Niamey

N'Djamena

Asmara

DJIBOUTI

Sanaa

SRI
LANKA

Phnom Penh

NIGERIA

SUDAN

Khartoum

Ho Chi
Minh City

BRUNEI

PALAU

FEDERATED STATES
OF MICRONESIA

ENIN

Abuja

ETHIOPIA

Addis Ababa

Colombo

Kuala
Lumpur

OGO

Lagos

CAMEROON

CENTRAL
AFRICAN
REPUBLIC

SOMALIA

MALDIVES

MALAYSIA

JATORIAL
NEA

DEMOCRATIC
REPUBLIC
OF THE
CONGO

UGANDA

Singapore

SINGAPORE

NAURU

KIRIBATI

GABON

REP
OF THE
CONGO

KENYA

Nairobi

INDONESIA

0° Equator

O TOMÉ
AND
INCIPE

RWANDA

BURUNDI

TANZANIA

SEYCHELLES

Cabinda
(ANGOLA)

Kinshasa

Dodoma

Dar es Salaam

PAPUA
NEW
GUINEA

Luanda

MALAWI

COMOROS

Surabaya

SOLOMON
ISLANDS

TUVALU

ANGOLA

ZAMBIA

MOZAMBIQUE

INDIAN OCEAN

Jakarta

Port Moresby

EAST TIMOR

Lusaka

MADAGASCAR

VANUATU

NAMIBIA

ZIMBABWE

Harare

Antananarivo

MAURITIUS

New Caledonia
(FRANCE)

FIJI

Windhoek

BOTSWANA

Réunion
(FRANCE)

20°S

Gaborone

Pretoria

AUSTRALIA

Tropic of Capricorn

Johannesburg

Maputo

SWAZILAND

Bloemfontein

LESOTHO

SOUTH
AFRICA

Cape Town

Sydney

Canberra

NEW
ZEALAND

Melbourne

20°E

40°E

60°E

80°E

100°E

120°E

140°E

160°E

60°S

Wellington

ANTARCTICA

Tasmania

	COUNTRY	CAPITAL
1	Czech Republic	Prague
2	Slovakia	Bratislava
3	Slovenia	Ljubljana
4	Croatia	Zagreb
5	Bosnia and Herzegovina	Sarajevo
6	Macedonia	Skopje
7	Serbia	Belgrade
8	Montenegro	Podgorica
9	Lithuania	Vilnius
10	Latvia	Riga
11	Estonia	Tallinn

ICELAND

Reykjavik

Arctic
Circle

SWEDEN

FINLAND

NORWAY

Helsinki

0 250 500 Miles
0 250 500 Kilometers
Projection: Mollweide

RUSSIA

60°N

Oslo

Stockholm

St. Petersburg

11

UNITED
KINGDOM

North
Sea

DENMARK

Copenhagen

10

Minsk

9

NETHERLANDS

Amsterdam

Berlin

Warsaw

BELARUS

Moscow

Dublin

London

Brussels

GERMANY

POLAND

Kiev

50°N

IRELAND

BELGIUM

Paris

LUXEMBOURG

1

Vienna

2

Budapest

UKRAINE

ATLANTIC
OCEAN

FRANCE

SWITZERLAND

Bern

AUSTRIA

HUNGARY

Chisinau

MOLDOVA

LIECHTENSTEIN

3

4

ROMANIA

Corsica
(FRANCE)

MONACO

ITALY

5

7

Bucharest

SAN MARINO

Rome

BULGARIA

Black Sea

ANDORRA

VATICAN CITY

8

6

Sofia

40°N

PORTUGAL

Balearic
Is.
(SPAIN)

Sardinia
(ITALY)

Tirane

ALBANIA

GREECE

Madrid

SPAIN

Lisbon

Mediterranean

Sicily

MALTA

Sea

Athens

Gibraltar (U.K.)

Crete

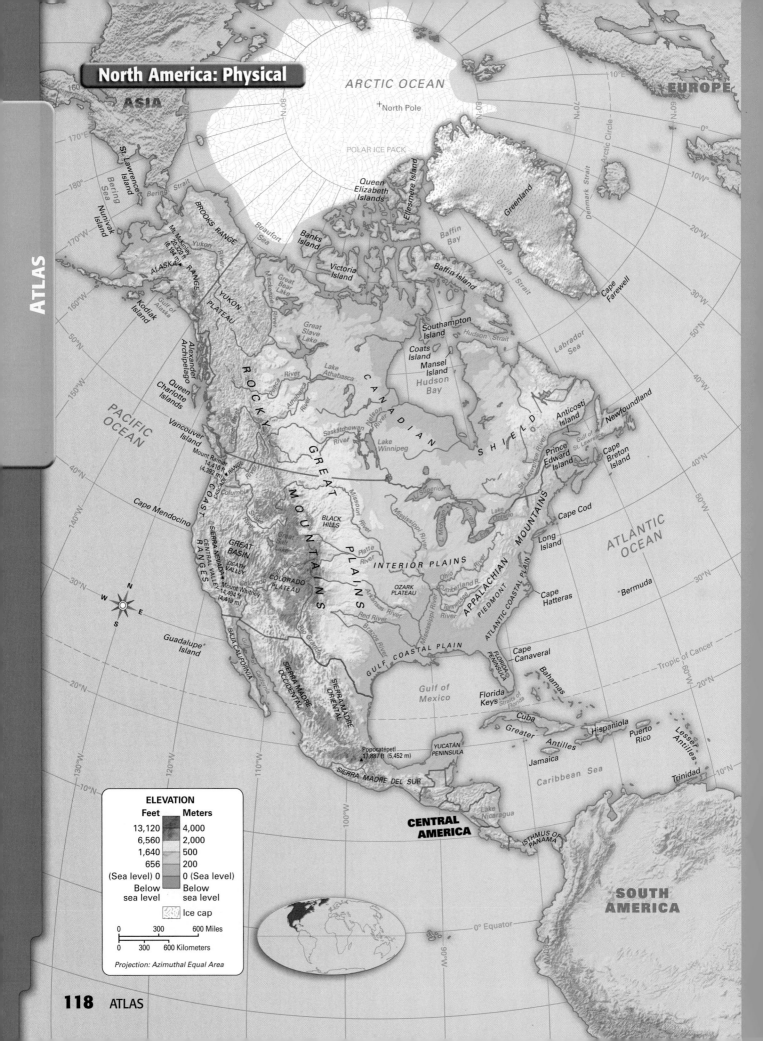

North America: Physical

ASIA

ARCTIC OCEAN

+ North Pole

POLAR ICE PACK

EUROPE

Queen
Elizabeth
Islands

Ellesmere Island

Greenland

St. Lawrence Island

BROOKS RANGE

Beaufort Sea

Banks Island

Baffin Bay

Denmark Strait

Bering Strait

Nunivak Island

Yukon River

Mt. McKinley
20,320 ft
(6,194 m)

ALASKA RANGE

YUKON PLATEAU

Great Bear Lake

Mackenzie River

Victoria Island

Baffin Island

Davis Strait

Cape Farewell

Arctic Circle

Kodiak Island

Gulf of Alaska

Alexander Archipelago

Queen Charlotte Islands

Vancouver Island

ROCKY

Peace River

Athabasca River

Great Slave Lake

Lake Athabasca

C A N A D I A N

Southampton Island

Coats Island

Mansel Island

Hudson Strait

Hudson Bay

Labrador Sea

PACIFIC OCEAN

Mount Rainier
14,410 ft
(4,392 m)

CASCADE RANGE

COAST RANGES

Columbia River

GREAT

Saskatchewan River

Lake Winnipeg

Nelson River

S H I E L D

Anticosti Island

Newfoundland

Prince Edward Island

Gulf of St. Lawrence

Cape Breton Island

Cape Mendocino

SIERRA NEVADA

GREAT BASIN

Great Salt Lake

DEATH VALLEY

BLACK HILLS

M O U N T A I N S

Missouri River

Lake Superior

Lake Michigan

Lake Huron

St. Lawrence River

Lake Ontario

A P P A L A C H I A N

Cape Cod

Long Island

ATLANTIC OCEAN

CENTRAL VALLEY

Mount Whitney
14,494 ft
(4,419 m)

COLORADO PLATEAU

Colorado River

P L A I N S

Platte River

INTERIOR PLAINS

OZARK PLATEAU

Ohio River

Cumberland R.

Tennessee River

M O U N T A I N S

PIEDMONT

Cape Hatteras

Bermuda

Guadalupe Island

BAJA CALIFORNIA

SIERRA MADRE OCCIDENTAL

Gulf of California

Rio Grande

Arkansas River

Red River

Brazos River

Mississippi River

GULF COASTAL PLAIN

ATLANTIC COASTAL PLAIN

FLORIDA PENINSULA

Cape Canaveral

Tropic of Cancer

SIERRA MADRE ORIENTAL

Gulf of Mexico

Florida Keys

Straits of Florida

Bahamas

Popocatépetl
17,887 ft (5,452 m)

YUCATÁN PENINSULA

Cuba

Greater Antilles

Jamaica

Hispaniola

Puerto Rico

Lesser Antilles

Caribbean Sea

Trinidad

SIERRA MADRE DEL SUR

CENTRAL AMERICA

Lake Nicaragua

ISTHMUS OF PANAMA

SOUTH AMERICA

0° Equator

ELEVATION

Feet	Meters
13,120	4,000
6,560	2,000
1,640	500
656	200
(Sea level) 0	0 (Sea level)
Below sea level	Below sea level

Ice cap

0   300   600 Miles

0   300   600 Kilometers

Projection: Azimuthal Equal Area

118 ATLAS

ATLAS

ARCTIC OCEAN

ASIA

EUROPE

North Pole

ATLAS

PACIFIC OCEAN

ALASKA (U.S.)

St. Lawrence Island

Nunivak Island

Bering Sea

Point Barrow

Beaufort Sea

Banks Island

Victoria Island

Queen Elizabeth Islands

Ellesmere Island

Baffin Bay

Greenland (DENMARK)

ICELAND

Denmark Strait

Arctic Circle

Cape Farewell

Anchorage

Kodiak Island

Gulf of Alaska

Alexander Archipelago

Juneau

Queen Charlotte Islands

Vancouver Island

Vancouver

Seattle

Portland

Edmonton

Calgary

Winnipeg

Great Bear Lake

Great Slave Lake

Lake Winnipeg

CANADA

Southampton Island

Coats Island

Mansel Island

Hudson Bay

Hudson Strait

Baffin Island

Davis Strait

Labrador Sea

Anticosti Island

Newfoundland

St. Pierre and Miquelon (FRANCE)

Cape Breton Island

Prince Edward Island

Gulf of St. Lawrence

Quebec

Montreal

Ottawa

Toronto

Lake Superior

Lake Michigan

Lake Huron

Lake Ontario

Lake Erie

Boston

Cape Cod

New York City

Philadelphia

Baltimore

Washington, D.C.

Norfolk

ATLANTIC OCEAN

San Francisco

San Jose

Salt Lake City

Great Salt Lake

Denver

Kansas City

Minneapolis

Milwaukee

Chicago

Indianapolis

St. Louis

Detroit

Cleveland

Columbus

UNITED STATES

Los Angeles

San Diego

Tijuana

Phoenix

Memphis

Dallas

Atlanta

Birmingham

Jacksonville

Bermuda (U.K.)

Gulf of California

Austin

San Antonio

Houston

New Orleans

Gulf of Mexico

Miami

Florida Keys

BAHAMAS

Nassau

Turks and Caicos Islands (U.K.)

Tropic of Cancer

Monterrey

Mérida

Havana

Straits of Florida

CUBA

Cayman Is. (U.K.)

Kingston

JAMAICA

Port-au-Prince

HAITI

DOMINICAN REPUBLIC

Santo Domingo

San Juan

Puerto Rico (U.S.)

Virgin Is. (U.S., U.K.)

ST. KITTS & NEVIS

ANTIGUA & BARBUDA

Guadeloupe (FRANCE)

DOMINICA

Martinique (FRANCE)

ST. LUCIA

BARBADOS

MEXICO

Guadalajara

Mexico City

Puebla

Belmopan

BELIZE

GUATEMALA

Guatemala City

San Salvador

EL SALVADOR

HONDURAS

Tegucigalpa

NICARAGUA

Managua

Caribbean Sea

Aruba (NETHERLANDS)

Netherlands Antilles (NETHERLANDS)

ST. VINCENT AND THE GRENADINES

GRENADA

TRINIDAD AND TOBAGO

San José

COSTA RICA

PANAMA

Panama City

Panama Canal

SOUTH AMERICA

Equator

Legend:
⊛ National capital
• Other city

0   300   600 Miles
0   300   600 Kilometers

Projection: Azimuthal Equal-Area

South America: Physical

CENTRAL AMERICA

Caribbean Sea

Panama Canal

Lake Maracaibo

Margarita Island
Tobago
Trinidad
Orinoco River Delta

Gulf of Panama

Malpelo Island

Mount Tolima
18,425 ft
(5,616 m)

Chocó River

Magdalena River

LLANOS

Meta River

Orinoco River

Angel Falls

GUIANA HIGHLANDS

Devil's Island
Cape Orange

Orinoco River

Amazon River Delta

Caqueta River

Japurá River

Río Negro

Amazon River

ATLANTIC OCEAN

Galápagos Islands

0° Equator

Gulf of Guayaquil

Marañón River

Mount Chimborazo
20,561 ft
(6,267 m)

AMAZON

BASIN

Amazon River

Tapajós River

Amazon River

Ucayali River

Juruá River

Purus

Madeira River

Tocantins River

Parnaíba River

River

ANDES

Mount Huascarán
22,205 ft
(6,768 m)

PACIFIC OCEAN

Beni River

Mamoré

MATO GROSSO PLATEAU

Xingu River

Araguaia River

BRAZILIAN HIGHLANDS

São Francisco River

Ancohuma Peak
20,958 ft
(6,388 m)

River

Lake Poopó

Lake Titicaca

ATACAMA DESERT

CHACO

São

River

Paraguay River

BRAZILIAN PLATEAU

San Félix Island
San Ambrosio Island

Tropic of Capricorn

Salado River

Paraná River

River

ANDES

Mount Aconcagua
22,834 ft
(6,960 m)

Salado River

Uruguay River

ATLANTIC OCEAN

Juan Fernández Islands

PAMPAS

Río de la Plata

Colorado River

PATAGONIA

Gulf of San Matias

Chiloé Island

Chonos Archipelago

Gulf of San Jorge

Cape Tres Puntas

Bahía Grande

Strait of Magellan

Falkland Islands

South Georgia Islands

Tierra del Fuego

Cape Horn

ELEVATION

Feet	Meters
13,120	4,000
6,560	2,000
1,640	500
656	200
(Sea level) 0	0 (Sea level)
Below sea level	Below sea level

0   250   500 Miles

0   250   500 Kilometers

Projection: Azimuthal Equal Area

CENTRAL
AMERICA

Caribbean Sea

Barranquilla
Cartagena

Caracas

Lake
Maracaibo

VENEZUELA

Medellín

Bogotá

COLOMBIA

Malpelo
Island
(COLOMBIA)

Cali

Georgetown
Paramaribo

GUYANA

Cayenne

SURINAME

French
Guiana
(FRANCE)

ATLANTIC
OCEAN

Quito

ECUADOR

Guayaquil

Galápagos
Islands
(ECUADOR)

Belém

0° Equator

PERU

Trujillo

BRAZIL

Recife

Callao
Lima

PACIFIC
OCEAN

Arequipa

Lake
Titicaca

La Paz

Lake
Poopó

BOLIVIA

Sucre

Brasília

Salvador

Belo Horizonte

20°S

PARAGUAY

Campinas
São Paulo

Asunción

Rio de Janeiro

Tropic of Capricorn

San Ambrosio
Island
(CHILE)

San Félix Island
(CHILE)

Curitiba

Pôrto Alegre

CHILE

Juan Fernández
Islands
(CHILE)

Córdoba

Valparaíso
Santiago

Rosario

URUGUAY

Buenos Aires

Montevideo

ATLANTIC
OCEAN

30°S

ARGENTINA

40°S

National capital

Other city

| 0 | 250 | 500 Miles |

| 0 | 250 | 500 Kilometers |

Projection: Azimuthal Equal-Area

Strait of
Magellan

Falkland
Islands (U.K.)

Tierra del
Fuego

South Georgia
Island
(U.K.)

ATLAS

ASIA

SOUTHWEST ASIA

AFRICA

URAL MOUNTAINS

Ural River

Kama River

Pechora River

Barents Sea

North Dvina River

White Sea

KOLA PENINSULA

Lake Onega

Lake Ladoga

Rybinsk Reservoir

Volga River

Don River

Sea of Azov

CAUCASUS MTS.

Mt. Elbrus 18,510 ft (5,642 m)

Caspian Sea

Black Sea

CRIMEAN PENINSULA

Dnipro River

Dniester River

Nistru River

NORTHERN EUROPEAN PLAIN

BALTIC PLAINS

Gulf of Finland

Daugava R.

Neman R.

Vistula River

Oder River

CARPATHIAN MTS.

TRANSYLVANIAN ALPS

Danube River

Tisza River

BALKAN PENINSULA

Aegean Sea

Sea of Marmara

Rhodes

Crete

Malta

Sicily

DINARIC ALPS

Adriatic Sea

APENNINES

Tiber River

Tyrrhenian Sea

Sardinia

Corsica

Balearic Islands

Mediterranean Sea

ARCTIC OCEAN

North Cape

KJÖLEN MOUNTAINS

Norwegian Sea

Gulf of Bothnia

Lake Vänern

Lake Vättern

Kattegat

Skagerrak

Baltic Sea

Elbe River

Rhine River

Danube River

ALPS

Lake Geneva

Mont Blanc 15,781 ft (4,810 m)

Rhône River

PYRENEES

Ebro River

IBERIAN PENINSULA

Duero River

Tagus River

Guadiana River

Guadalquivir River

Strait of Gibraltar

Cape Finisterre

Bay of Biscay

Garonne River

Loire River

Seine River

English Channel

Thames River

PENNINES

British Isles

Irish Sea

Hebrides

Orkney Islands

Shetland Islands

Faeroe Islands

Iceland

North Sea

Arctic Circle

ATLANTIC OCEAN

N E S W

70°N

60°N

50°N

40°N

30°N

20°N

40°W

30°W

20°W

10°W

0°

10°E

20°E

30°E

40°E

50°E

60°E

70°N

Projection: Azimuthal Equal Area

Europe: Physical

ELEVATION

Feet	Meters
13,120	4,000
6,560	2,000
1,640	500
656	200
0 (Sea level)	0 (Sea level)
Below sea level	Below sea level

Ice cap

300 Miles
0   150   300 Kilometers
0   150

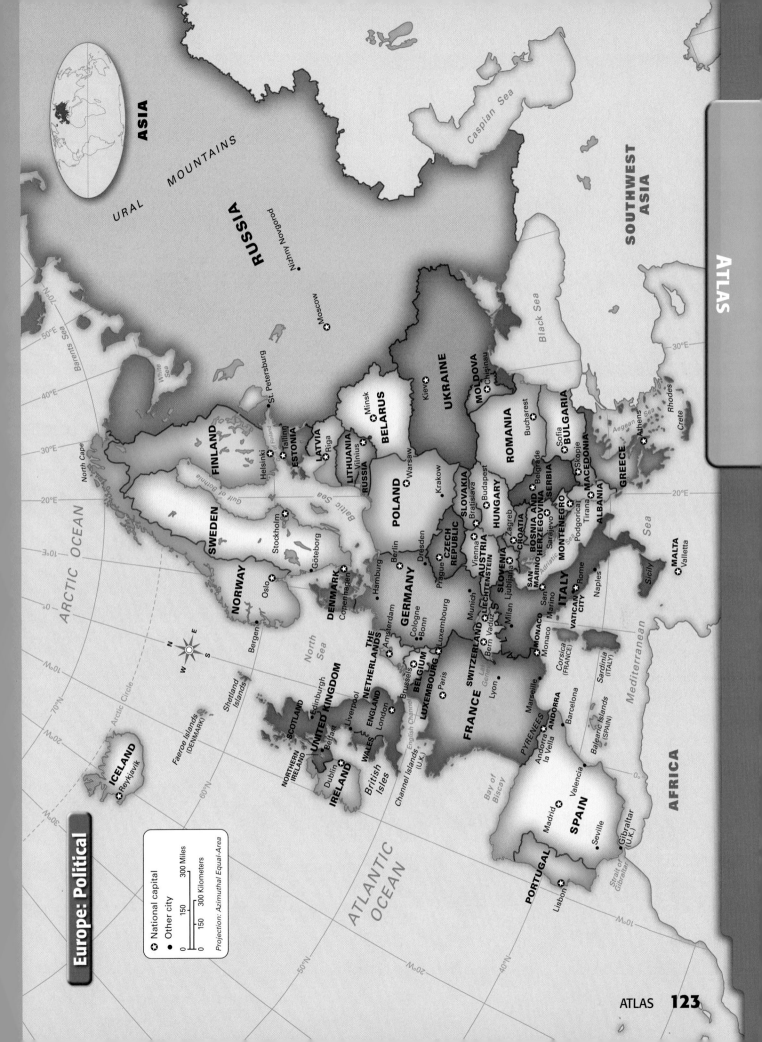

Europe: Political

ATLAS

Legend:
- ✪ National capital
- • Other city

0   150   300 Miles
0   150   300 Kilometers

Projection: Azimuthal Equal-Area

ASIA

URAL MOUNTAINS

RUSSIA

Nizhny Novgorod

Moscow ✪

SOUTHWEST ASIA

Caspian Sea

Black Sea

St. Petersburg

FINLAND

Helsinki ✪

Gulf of Bothnia

ESTONIA
Tallinn ✪

LATVIA
Riga ✪

LITHUANIA
Vilnius ✪

RUSSIA

BELARUS
Minsk ✪

UKRAINE
Kiev ✪

MOLDOVA
Chişinău ✪

ROMANIA
Bucharest ✪

BULGARIA
Sofia ✪

Barents Sea

White Sea

North Cape

SWEDEN

Stockholm ✪

Göteborg •

POLAND
Warsaw ✪
Kraków •

SLOVAKIA
Bratislava ✪

HUNGARY
Budapest ✪

SERBIA
Belgrade ✪

MACEDONIA
Skopje ✪

ALBANIA
Tirana ✪

GREECE
Athens ✪

Aegean Sea

Rhodes

Crete

ARCTIC OCEAN

NORWAY
Oslo ✪
Bergen •

DENMARK
Copenhagen ✪

Hamburg •

GERMANY
Berlin ✪
Dresden •
Cologne • Bonn •

Baltic Sea

CZECH REPUBLIC
Prague ✪

AUSTRIA
Vienna ✪

SLOVENIA
Ljubljana ✪

CROATIA
Zagreb ✪

BOSNIA AND HERZEGOVINA
Sarajevo ✪

MONTENEGRO
Podgorica ✪

SAN MARINO
San Marino ✪

ITALY
Rome ✪
Naples •

VATICAN CITY

Adriatic Sea

Sicily

MALTA
Valletta ✪

North Sea

THE NETHERLANDS
Amsterdam ✪

BELGIUM
Brussels ✪

LUXEMBOURG
Luxembourg ✪

SWITZERLAND
Bern ✪

LIECHTENSTEIN
Vaduz ✪

Munich •

Lake Geneva

Milan •

MONACO
Monaco ✪

Corsica (FRANCE)

Sardinia (ITALY)

Mediterranean Sea

ICELAND
Reykjavik ✪

Faeroe Islands (DENMARK)

Shetland Islands

SCOTLAND
Edinburgh •

NORTHERN IRELAND
Belfast •

IRELAND
Dublin ✪

UNITED KINGDOM
Liverpool •

WALES

ENGLAND
London ✪

British Isles

Channel Islands (U.K.)

English Channel

FRANCE
Paris ✪
Lyon •
Marseille •

PYRENEES

ANDORRA
Andorra la Vella ✪

Bay of Biscay

SPAIN
Madrid ✪
Valencia •
Barcelona •
Seville •

Gibraltar (U.K.)

Strait of Gibraltar

Balearic Islands (SPAIN)

PORTUGAL
Lisbon ✪

ATLANTIC OCEAN

AFRICA

Arctic Circle

Arctic Circle

N E S W

Asia: Physical

ELEVATION

Feet	Meters
13,120	4,000
6,560	2,000
1,640	500
656	200
(Sea level) 0	0 (Sea level)
Below sea level	Below sea level

Ice cap

750 Miles
0 250 500 750 Kilometers

Projection: Two-Point Equidistant

EUROPE
AFRICA
AUSTRALIA

PACIFIC OCEAN
INDIAN OCEAN

Equator
Tropic of Cancer

North Pole
Arctic Circle

North Land
Franz Josef Land
Novaya Zemlya
Wrangel Island
New Siberian Islands
Taymyr Peninsula

Barents Sea
Kara Sea
Laptev Sea
Bering Sea

Aleutian Islands
Kamchatka Peninsula
Sea of Okhotsk
Sakhalin Island
Kuril Islands
Hokkaido
Honshu
Shikoku
Kyushu
Sea of Japan (East Sea)
Korea Strait
Okinawa
Ryukyu Islands
Taiwan
East China Sea
Yellow Sea
Luzon Strait
Luzon
Philippines
Mindanao
Celebes Sea
Celebes
Molucca B.
Banda Sea
Arafura Sea
New Guinea
Maoke Mountains
Borneo
Java Sea
Java
Bangka
Sumatra
Mentawai Islands
Malay Peninsula
Gulf of Thailand
South China Sea
Hainan
Gulf of Tonkin
Indochina Peninsula
Mekong River
Chao Phraya River
Andaman Sea
Andaman Islands
Nicobar Islands
Bay of Bengal
Sri Lanka
Maldives
Lakshadweep Islands
Socotra Island
Arabian Sea
Gulf of Oman
Gulf of Aden
Red Sea
Persian Gulf

SIBERIA
Central Siberian Plateau
West Siberian Plain
Kazakh Uplands
Ural Mountains
Ob River
Tunguska River
Yenisey River
Angara River
Lower Tunguska River
Lena River
Aldan River
Shilka River
Amur River
Kolyma Mts.
Chersky Range
Verkhoyansk Range
Yablonovy Range
Stanovoy Mountains
Central Range
Sayan Mountains
Altay Mountains
Mongolian Plateau
Greater Khingan Range
Gobi
North China Plain
Qin Ling
Yellow River (Huang He)
Chang (Yangtze) River
Xi River
Bohai
Tarim Basin
Taklimakan Desert
Tian Shan
Kunlun Mountains
Plateau of Tibet
Himalayas
Mount Everest 29,035 ft (8,850 m)
Indo-Gangetic Plain
Ganges River
Brahmaputra River
Irrawaddy River
Salween River
Sutlej River
Indus River
Thar Desert
Deccan Plateau
Eastern Ghats
Western Ghats
Godavari River
Hindu Kush
Kara Kum
Kyzyl Kum
Turan Lowland
Amu Darya
Syr Darya
Aral Sea
Balqash Lake
Irtysh River
Ishim River
Ob River
Ural River
Ustyurt Plateau
Caspian Sea
Great Salt Desert
Zagros Mts.
Caucasus Mts.
Mount Ararat 16,945 ft (5,165 m)
Anatolian Plateau
Tigris River
Euphrates River
Syrian Desert
An-Nafud
Rub' Al-Khali
Sinai Peninsula
Cyprus
Mediterranean Sea
Black Sea
Bosporus

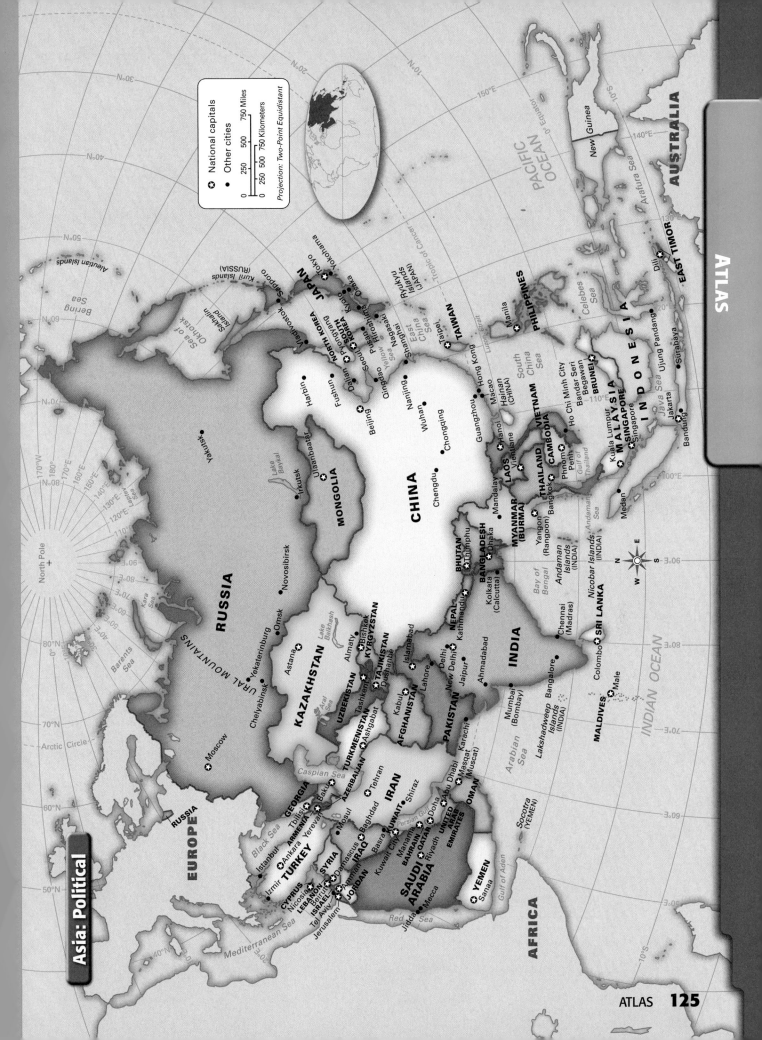

Asia: Political

National capitals
Other cities

| 0 | 250 | 500 | 750 Miles |
| 0 | 250 | 500 | 750 Kilometers |

Projection: Two-Point Equidistant

PACIFIC OCEAN

AUSTRALIA

New Guinea

EAST TIMOR
Dili

Arafura Sea

Celebes Sea

PHILIPPINES
Manila

INDONESIA

BRUNEI
Bandar Seri Begawan

MALAYSIA
SINGAPORE
Singapore
Kuala Lumpur

Ujung Pandang
Surabaya
Jakarta
Java Sea
Bandung

Medan

South China Sea

VIETNAM
Ho Chi Minh City

CAMBODIA
Phnom Penh

Gulf of Thailand

Hainan (CHINA)

Macao
Hong Kong
Guangzhou

TAIWAN
Taipei

East China Sea

Ryukyu Islands (JAPAN)

Nagasaki
Hiroshima
Kyoto
Osaka
Yokohama
Tokyo

JAPAN

Sapporo

Kuril Islands (RUSSIA)

Sakhalin Island

Sea of Okhotsk

Vladivostok

Aleutian Islands

Bering Sea

North Pole

Arctic Circle

RUSSIA

URAL MOUNTAINS

EUROPE

RUSSIA

Moscow

Yekaterinburg
Chelyabinsk
Omsk
Novosibirsk

Yakutsk

Irkutsk

Lake Baykal

Ulaanbaatar

MONGOLIA

CHINA

Chengdu
Chongqing
Wuhan

Nanjing
Shanghai

Beijing

Yellow Sea

Qingdao
Dalian
SOUTH KOREA
Seoul
Pusan

NORTH KOREA
Pyongyang

Fushun
Harbin

Astana

KAZAKHSTAN

Aral Sea

Lake Balkhash

Almaty
Bishkek
KYRGYZSTAN
Tashkent
UZBEKISTAN
TAJIKISTAN
Dushanbe
Ashgabat
TURKMENISTAN

Kabul
AFGHANISTAN

Kathmandu
NEPAL
BHUTAN
Thimphu

Islamabad

Lahore
PAKISTAN
Delhi
New Delhi
Jaipur
Ahmadabad

Karachi

BANGLADESH
Dhaka

Kolkata (Calcutta)

MYANMAR (BURMA)
Mandalay
Yangon (Rangoon)

LAOS
Vientiane

THAILAND
Bangkok

Bay of Bengal

Andaman Islands (INDIA)

Andaman Sea

Nicobar Islands (INDIA)

INDIA

Mumbai (Bombay)

Bangalore
Chennai (Madras)

Lakshadweep Islands (INDIA)

SRI LANKA
Colombo

MALDIVES
Male

Arabian Sea

INDIAN OCEAN

Caspian Sea

GEORGIA
Tbilisi
ARMENIA
Yerevan
AZERBAIJAN
Baku

Tehran

IRAN

Shiraz

Mosul
Baghdad
IRAQ
Basra

KUWAIT
Kuwait City

Persian Gulf

BAHRAIN
Manama
QATAR
Doha
UNITED ARAB EMIRATES
Abu Dhabi

OMAN
Masqat (Muscat)

Socotra (YEMEN)

Gulf of Aden

YEMEN
Sanaa

SAUDI ARABIA
Riyadh

Mecca
Jidda

Red Sea

AFRICA

JORDAN
Amman

ISRAEL
Tel Aviv
Jerusalem

LEBANON
Beirut
SYRIA
Damascus

CYPRUS
Nicosia

TURKEY
Ankara
Istanbul
Izmir

Black Sea

Mediterranean Sea

Tropic of Cancer

Equator

Africa: Physical

EUROPE

SOUTHWEST
ASIA

Azores

Madeira
Islands

Strait of
Gibraltar

Mediterranean Sea

Gulf of
Sidra

QATTARA
DEPRESSION

Suez Canal

Persian Gulf

30°N

ATLAS MOUNTAINS

Canary
Islands

S A H A R A

LIBYAN DESERT

Tropic of Cancer

20°N

Cape
Blanc

EL DJOUF

AHAGGAR
MOUNTAINS

TIBESTI
MOUNTAINS

Nile River

Lake
Nasser

NUBIAN
DESERT

Red Sea

Cape Verde
Islands

AIR MTS.

S A H E L

Lake
Tana

Gulf of Aden

10°N

Cape
Verde

Niger River

S U D A N

CHAD
BASIN

Lake
Chad

Blue Nile

White Nile

ETHIOPIAN
HIGHLANDS

HORN OF AFRICA

SOMALI
PENINSULA

10°N

FOUTA
DJALLON

Senegal R.

White Volta R.

Black Volta R.

Benue River

SUDAN
BASIN

Lake
Volta

ADAMAWA MTS.

Ubangi River

Lake
Turkana

RIFT VALLEY

Mount Kenya
17,058 ft
(5,199 m)

Cape
Palmas

Gulf of
Guinea

Congo River

CONGO
BASIN

Lake
Albert

Lake
Edward

Lake
Victoria

SERENGETI
PLAIN

MASAI
STEPPE

Mount Kilimanjaro
19,340 ft
(5,895 m)

INDIAN
OCEAN

0° Equator

Cape
Lopez

Kasai River

Lake
Kivu

Lake
Tanganyika

EASTERN RIFT

Zanzibar

0° Equator

N
W E
S

MITUMBA MOUNTAINS

WESTERN RIFT VALLEY

Lake
Rukwa

Seychelles

Ascension

ATLANTIC
OCEAN

Cuanza River

Lake
Mweru

Lake Malawi
(Nyasa)

Cape Delgado

10°S

Comoro
Islands

10°S

Lake
Kariba

Zambezi River

Mozambique Channel

Madagascar

Mauritius

Okavango
Delta

Victoria
Falls

Réunion

20°S

NAMIB DESERT

KALAHARI BASIN

KALAHARI
DESERT

Limpopo River

Tropic of Capricorn

Tropic of Capricorn

Orange River

Vaal River

DRAKENSBERG MOUNTAINS

30°S

GREAT
KARROO

Cape of
Good Hope

30°S

ELEVATION

Feet	Meters
13,120	4,000
6,560	2,000
1,640	500
656	200
(Sea level) 0	0 (Sea level)
Below sea level	Below sea level

0   250   500 Miles

0   250   500 Kilometers

Projection: Azimuthal Equal-Area

EUROPE

SOUTHWEST
ASIA

ATLAS

Azores
(PORTUGAL)

Madeira
(PORTUGAL)

Strait of
Gibraltar

Algiers Tunis

Casablanca Rabat

TUNISIA

Tripoli

Mediterranean Sea

Alexandria

Giza Cairo

MOROCCO

Canary Islands
(SPAIN)

El Aaiún

WESTERN
SAHARA
(Claimed by
Morocco)

ALGERIA

LIBYA

EGYPT

Tropic of Cancer

CAPE
VERDE

Praia

MAURITANIA

Nouakchott

MALI

SENEGAL

Dakar

GAMBIA

Banjul

Bamako

Bissau

GUINEA
BISSAU

GUINEA

Conakry

Freetown

SIERRA LEONE

Monrovia

LIBERIA

NIGER

Niamey

BURKINA
FASO

Ouagadougou

CÔTE
D'IVOIRE

GHANA

Yamoussoukro

Abidjan

Accra

BENIN

TOGO

Lomé

NIGERIA

Abuja

Lagos

Porto
Novo

CHAD

Lake
Chad

N'Djamena

Khartoum

SUDAN

ERITREA

Asmara

DJIBOUTI

Djibouti

ETHIOPIA

Addis Ababa

Red Sea

Gulf of Aden

10°N

Malabo

EQUATORIAL GUINEA

SÃO TOMÉ AND PRÍNCIPE

São Tomé

CAMEROON

Yaoundé

Gulf of
Guinea

Bangui

CENTRAL AFRICAN
REPUBLIC

Libreville

REPUBLIC
OF THE
CONGO

GABON

Kisangani

UGANDA

Kampala

KENYA

Nairobi

SOMALIA

Mogadishu

0° Equator

Brazzaville

CABINDA
(ANGOLA)

Kinshasa

DEMOCRATIC
REPUBLIC
OF THE CONGO

Bujumbura

RWANDA

Kigali

BURUNDI

TANZANIA

Dodoma

Dar es Salaam

Mombasa

Pemba

Zanzibar

INDIAN
OCEAN

Victoria

SEYCHELLES

Luanda

N

W E

S

ATLANTIC
OCEAN

Lubumbashi

ANGOLA

ZAMBIA

Lusaka

MALAWI

Lilongwe

COMOROS

Moroni

St. Helena
(U.K.)

Harare

ZIMBABWE

Bulawayo

MOZAMBIQUE

MADAGASCAR

Antananarivo

MAURITIUS

Port Louis

Réunion
(FRANCE)

Tropic of Capricorn

NAMIBIA

Windhoek

BOTSWANA

Gaborone

Pretoria

Maputo

Johannesburg

Mbabane

Bloemfontein

SWAZILAND

Maseru

LESOTHO

SOUTH AFRICA

Cape Town

✪ National capital
● Other city

0   250   500 Miles
0   250   500 Kilometers

Projection: Azimuthal Equal-Area

The Pacific: Political

NORTH AMERICA

ASIA

NORTH PACIFIC OCEAN

SOUTH PACIFIC OCEAN

AUSTRALIA

Tropic of Cancer

Tropic of Capricorn

0° Equator

International Date Line

Philippine Sea

South China Sea

Timor Sea

Arafura Sea

Coral Sea

Tasman Sea

INDIAN OCEAN

MICRONESIA

MELANESIA

POLYNESIA

Legend
- ✪ National capital
- ● Other city

1,000 Miles
1,000 Kilometers
500

Projection: Azimuthal Equal-Area

Easter Island (CHILE)

Pitcairn (U.K.)
Pitcairn Island
Ducie Island

Marquesas Islands (FRANCE)
Tuamotu Archipelago (FRANCE)
Rapa Island (FRANCE)
Tubuai Islands (FRANCE)
French Polynesia
Society Islands (FRANCE)
Tahiti (FRANCE)
Papeete

Starbuck Island

Manihiki Island
Cook Islands (NEW ZEALAND)
Rarotonga Island

Hawaiian Islands
Hawaii (U.S.)

Kingman Reef (U.S.)
Palmyra Island (U.S.)
Fanning Island
Washington Island
Jarvis I. (U.S.)

KIRIBATI

Phoenix Islands

American Samoa
Pago Pago
Niue (N.Z.)
SAMOA
Apia
TONGA
Nuku'alofa

Tokelau (N.Z.)
Howland I. (U.S.)
Baker I. (U.S.)
McKean I.
Gardner Island

Midway Island (U.S.)

Johnston Island (U.S.)

Wallis & Futuna (FR.)
TUVALU
Funafuti
FIJI
Suva

Wake Island (U.S.)

MARSHALL ISLANDS
Kwajalein Island
Majuro
Eniwetok I.

Tarawa
Gilbert Islands

NAURU
SOLOMON ISLANDS
Honiara
Guadalcanal I.
Bismarck Archipelago

Palikir
Truk Is.
FEDERATED STATES OF MICRONESIA

VANUATU
Espíritu Santo I.
Malekula I.
Port-Vila
New Caledonia (FRANCE)
Nouméa
Loyalty Islands (FRANCE)

Norfolk Island (AUSTRALIA)

Kermadec Islands (N.Z.)

Chatham Islands (N.Z.)
Bounty Islands (N.Z.)
Auckland Islands (NEW ZEALAND)

NEW ZEALAND
Auckland
Wellington
Christchurch
North Island
South Island

PAPUA NEW GUINEA
Port Moresby
New Guinea

PALAU
Koror

Northern Marianas (U.S.)
Guam (U.S.)
Agana

Bonin Islands (JAPAN)
Volcano Islands (JAPAN)

Brisbane
Sydney
Canberra
Hobart
Melbourne
Adelaide
Perth
Darwin

Christmas Island (AUSTRALIA)

30°N
15°N
15°S
30°S
45°S

135°W
150°W
165°W
180°
165°E
150°E
135°E
120°W
120°E

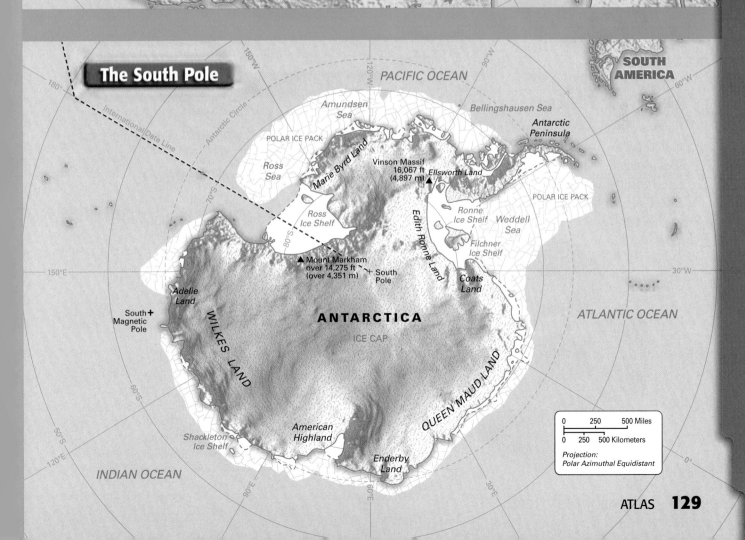

The North Pole

0   200   400 Miles
0   200   400 Kilometers

Projection:
Polar Azimuthal Equidistant

Kara
Sea

Barents
Sea

EUROPE

Norwegian
Sea

Arctic Circle

90°E

60°E

30°E

Laptev
Sea

120°E

ASIA

150°E

80°N

*ARCTIC
OCEAN*

North
Pole

Greenland
Sea

30°W

Greenland
(DENMARK)

**ATLANTIC
OCEAN**

60°N

International Date Line

POLAR ICE PACK

North
Magnetic
Pole

150°W

Baffin
Bay

60°W

Beaufort
Sea

180°

Bering Sea

50°N

**NORTH
AMERICA**

ATLAS

The South Pole

150°W

International Date Line

180°

Antarctic Circle

120°W

PACIFIC OCEAN

90°W

**SOUTH
AMERICA**

60°W

Amundsen
Sea

Bellingshausen Sea

POLAR ICE PACK

Ross
Sea

Marie Byrd Land

Vinson Massif
16,067 ft
(4,897 m)

Antarctic
Peninsula

Ellsworth Land

70°S

Ross
Ice Shelf

Ronne
Ice Shelf

Weddell
Sea

POLAR ICE PACK

80°S

Mount Markham
over 14,275 ft
(over 4,351 m)

South
Pole

Edith Ronne Land

Filchner
Ice Shelf

30°W

150°E

South
Magnetic
Pole

Adelie
Land

WILKES LAND

ANTARCTICA

ICE CAP

Coats
Land

ATLANTIC OCEAN

QUEEN MAUD LAND

60°S

American
Highland

50°S

Shackleton
Ice Shelf

120°E

INDIAN OCEAN

Enderby
Land

90°E

60°E

30°E

0   250   500 Miles
0   250   500 Kilometers

Projection:
Polar Azimuthal Equidistant

The Physical World

Inside the Earth

Earth's interior has several different layers. Deep inside the planet is the core. The inner core is solid, and the outer core is liquid. Above the core is the mantle, which is mostly solid rock with a molten layer on top. The surface layer of Earth includes the crust, which is made up of rocks and soil. Finally, the atmosphere extends from the crust into space. It supports much of the life on Earth.

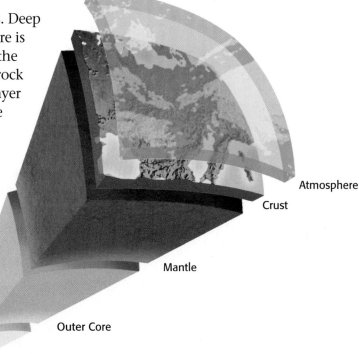

Atmosphere

Crust

Mantle

Outer Core

Inner Core

Tectonic Plates

Earth's crust is divided into huge pieces called tectonic plates, which fit together like a puzzle. As these plates slowly move, they collide and break apart, forming surface features like mountains, ocean basins, and ocean trenches.

Earth Facts	
Age:	4.6 billion years
Mass:	5,974,000,000,000,000,000,000 metric tons
Distance around the equator:	24,902 miles (40,067 km)
Distance around the poles:	24,860 miles (40,000 km)
Distance from the sun:	about 93 million miles (150 million km)
Earth's speed around the sun:	18.5 miles a second (29.8 km a second)
Percent of Earth's surface covered by water:	71%
What makes Earth unique:	large amounts of liquid water, tectonic activity, and life

The Continents

Geographers identify seven large landmasses, or continents, on Earth. Most of these continents are almost completely surrounded by water. Europe and Asia, however, are not. They share a long land boundary.

The world's continents are very different. For example, much of Australia is dry and rocky, while Antarctica is cold and icy. The information below highlights some key facts about each continent.

North America
- Percent of Earth's land: 16.5%
- Percent of Earth's population: 5.1%
- Lowest point: Death Valley, 282 feet (86 m) below sea level

South America
- Percent of Earth's land: 12%
- Percent of Earth's population: 8.6%
- Longest mountains: Andes, 4,500 miles (7,240 km)

Asia
- Percent of Earth's land: 30%
- Percent of Earth's population: 60.7%
- Highest point: Mount Everest, 29,035 feet (8,850 m)

Europe
- Percent of Earth's land: 6.7%
- Percent of Earth's population: 11.5%
- People per square mile: 187

Africa
- Percent of Earth's land: 20.2%
- Percent of Earth's population: 13.6%
- Longest river: Nile River, 4,160 miles (6,693 km)

Australia
- Percent of Earth's land: 5.2%
- Percent of Earth's population: 0.3%
- Oldest rocks: 3.7 billion years

Antarctica
- Percent of Earth's land: 8.9%
- Percent of Earth's population: 0%
- Coldest place: Plateau Station, -56.7°C (-70.1°F) average temperature

The Human World

World Population

More than 6 billion people live in the world today, and that number is growing quickly. Some people predict the world's population will reach 9 billion by 2050. As our population grows, it is also becoming more urban. Soon, as many people will live in cities and in towns as live in rural areas.

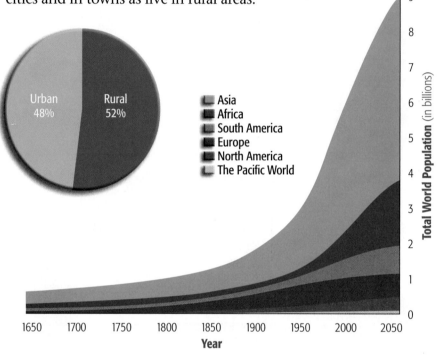

Urban 48%
Rural 52%

Asia
Africa
South America
Europe
North America
The Pacific World

Total World Population (in billions)

Year
1650 1700 1750 1800 1850 1900 1950 2000 2050

As the world's population grows, people are moving to already large cities such as Shanghai (above) and Hong Kong (right) in China.

Geographers divide the world into developed and less developed regions. In general, developed countries are wealthier and more urban, have lower population growth rates, and have higher life expectancies. As you can imagine, life is very different in developed and less developed regions.

Developed and Less Developed Countries

	Population	Rate of Natural Increase	Life Expectancy	Percent Urban	Per Capita GNP (U.S. $)
Developed Countries	1.2 billion	0.1%	76	76%	$23,690
Less Developed Countries	5.2 billion	1.5%	65	41%	$3,850
The World	6.4 billion	1.3%	67	48%	$7,590

World Religions

A large percentage of the world's people follow one of several major world religions. Christianity is the largest religion. About 33 percent of the world's people are Christian. Islam is the second-largest religion with about 20 percent. It is also the fastest-growing religion. Hinduism and Buddhism are also major world religions.

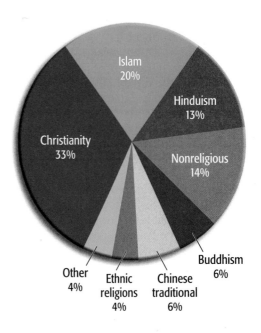

Islam 20%

Hinduism 13%

Christianity 33%

Nonreligious 14%

Buddhism 6%

Other 4%

Ethnic religions 4%

Chinese traditional 6%

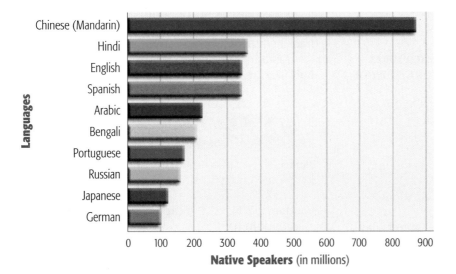

Languages

Chinese (Mandarin)
Hindi
English
Spanish
Arabic
Bengali
Portuguese
Russian
Japanese
German

0 100 200 300 400 500 600 700 800 900

Native Speakers (in millions)

World Languages

Although several thousand languages are spoken today, a handful of major languages have the largest numbers of native speakers. Chinese (Mandarin) is spoken by nearly one in six people. Hindi, English, Spanish, and Arabic are next, with native speakers all over the world.

Gazetteer

A

Afghanistan a landlocked country in Central Asia (p. 125)

Africa the second-largest continent; surrounded by the Atlantic Ocean, Indian Ocean, and Mediterranean Sea (p. 126)

Amazon River a major river in South America (p. 120)

Andes a mountain range along the west coast of South America (p. 120)

Antarctica the continent around the South Pole (p. 129)

Antarctic Circle the line of latitude located at 66.5° south of the equator; parallel beyond which no sunlight shines on the June solstice (p. 129)

Arctic Circle the line of latitude located at 66.5° north of the equator; parallel beyond which no sunlight shines on the December solstice (p. 129)

Arctic Ocean the ocean north of the Arctic Circle; the world's fourth-largest ocean (p. 114)

Asia the world's largest continent; located between Europe and the Pacific Ocean (p. 124)

Atlantic Ocean the ocean between the continents of North and South America and the continents of Europe and Africa; the world's second-largest ocean (p. 114)

Australia the only country occupying an entire continent (also called Australia); located between the Indian Ocean and the Pacific Ocean (p. 128)

C

California a state on the west coast of the United States (p. 112)

Canada the country north of the United States that occupies most of northern North America (p. 119)

Caribbean Sea an arm of the Atlantic Ocean between North and South America (p. 118)

Chicago (42°N, 88°W) a major city on Lake Michigan in northern Illinois (p. 113)

China a country in East Asia; the most populous country in the world (p. 125)

Colorado a state in the western United States (p. 112)

Cuba a Communist country and the largest island in the Caribbean Sea (p. 119)

E, F

Eastern Hemisphere the half of the globe between the prime meridian and 180° longitude that includes most of Africa and Europe as well as Asia, Australia, and the Indian Ocean (p. H7)

Egypt a country in North Africa between Libya and the Red Sea (p. 127)

equator the imaginary line of latitude that circles the globe halfway between the North and South Poles (p. H6)

Euphrates River a river in Southwest Asia (p. 124)

Europe the continent between the Ural Mountains and the Atlantic Ocean (p. 122)

Florida a state in the southern United States bordered mostly by the Atlantic Ocean and Gulf of Mexico (p. 113)

G, H, I

Grand Canyon a large canyon in the southwestern United States (p. 110)

Great Lakes the largest freshwater lake system in the world; located in North America (p. 111)

Great Plains a grassland region in the central United States (p. 111)

Gulf Stream a warm ocean current that flows north along the east coast of the United States (p. 52)

Hawaii an island state of the United States located in the Pacific Ocean (p. 112)

Himalayas a mountain system in Asia; the world's highest mountains (p. 124)

India a country in South Asia (p. 125)

Indian Ocean the world's third-largest ocean; located east of Africa, south of Asia, and west of Australia (p. 115)

Iraq a country located between Iran and Saudi Arabia in Southwest Asia (p. 125)

Italy a country in Southern Europe (p. 123)

J, K, L

Japan an island country in East Asia (p. 125)

Kenya a country in East Africa, south of Ethiopia (p. 127)

London (52°N, 0°) the capital of the United Kingdom (p. 123)

Mariana Trench the world's deepest ocean trench; located in the Pacific Ocean (p. 37)

Mediterranean Sea a sea surrounded by Europe, Asia, and Africa (p. 115)

Mexico a country in North America, south of the United States (p. 119)

Michigan a state in the north-central region of the United States (p. 113)

Mid-Atlantic Ridge a mid-ocean ridge located in the Atlantic Ocean (p. 37)

Mississippi River a major river in the central United States (p. 111)

Mount Saint Helens (46°N, 122°W) a volcano in Washington State; it erupted in 1980 (p. 43)

New York (41°N, 74°W) a city in the northeastern United States; the largest city in the United States (p. 113)

Nile River the world's longest river; flows through East Africa, Egypt, and into the Mediterranean Sea (p. 126)

North America a continent including Canada, the United States, Mexico, Central America, and the Caribbean islands (p. 119)

North Atlantic Current a warm ocean current that flows across the Atlantic Ocean and along Western Europe (p. 52)

Northern Hemisphere the northern half of the globe, between the equator and the North Pole (p. H7)

North Pole (90°N) the northern point of Earth's axis (p. 129)

Pacific Ocean the world's largest ocean; located between Asia and the Americas (p. 115)

prime meridian an imaginary line that runs through Greenwich, England, at 0° longitude (p. H6)

Ring of Fire a region that circles the Pacific Ocean; known for its earthquakes and volcanoes (p. 42)

Rocky Mountains a major mountain range in western North America (p. 110)

Sahara a desert in northern Africa; it is the largest desert in the world (p. 126)

San Andreas Fault an area at the boundary of the Pacific tectonic plate where earthquakes are common; located in California (p. 38)

Saudi Arabia a country occupying much of the Arabian Peninsula in Southwest Asia (p. 125)

Scandinavia a region of islands and peninsulas in far northern Europe; includes the countries of Norway, Sweden, Finland, and Denmark (p. 123)

South America a continent in the Western and Southern hemispheres (p. 121)

Southern Hemisphere the southern half of the globe, between the equator and the South Pole (p. H7)

South Pole (90°S) the southern point of Earth's axis (p. 129)

Texas a state in the southern United States (p. 113)

Tigris River a river in Southwest Asia (p. 124)

Tropic of Cancer the parallel 23.5° north of the equator; parallel on the globe at which the sun's most direct rays strike Earth during the June solstice (p. 28)

Tropic of Capricorn the parallel at 23.5° south of the equator; parallel on the globe at which the sun's most direct rays strike Earth during the December solstice (p. 28)

United States a country in North America located between Canada and Mexico (p. 112)

Washington a state in the northwestern United States (p. 112)

Washington, D.C. (39°N, 77°W) the capital of the United States (p. 113)

Western Hemisphere the half of the globe between 180° and the prime meridian that includes North and South America and the Pacific and Atlantic oceans (p. H7)

GAZETTEER

English and Spanish Glossary

MARK	AS IN	RESPELLING	EXAMPLE
a	alphabet	a	*AL-fuh-bet
ā	Asia	ay	AY-zhuh
ä	cart, top	ah	KAHRT, TAHP
e	let, ten	e	LET, TEN
ē	even, leaf	ee	EE-vuhn, LEEF
i	it, tip, British	i	IT, TIP, BRIT-ish
ī	site, buy, Ohio	y	SYT, BY, oh-HY-oh
	iris	eye	EYE-ris
k	card	k	KAHRD
kw	quest	kw	KWEST
ō	over, rainbow	oh	OH-vuhr, RAYN-boh
ù	book, wood	ooh	BOOHK, WOOHD
ò	all, orchid	aw	AWL, AWR-kid
òi	foil, coin	oy	FOYL, KOYN
aù	out	ow	OWT
ə	cup, butter	uh	KUHP, BUHT-uhr
ü	rule, food	oo	ROOL, FOOD
yü	few	yoo	FYOO
zh	vision	zh	VIZH-uhn

*A syllable printed in small capital letters receives heavier emphasis than the other syllable(s) in a word.

Phonetic Respelling and Pronunciation Guide

Many of the key terms in this textbook have been respelled to help you pronounce them. The letter combinations used in the respelling throughout the narrative are explained in this phonetic respelling and pronunciation guide. The guide is adapted from *Merriam-Webster's Collegiate Dictionary, Eleventh Edition; Merriam-Webster's Geographical Dictionary;* and *Merriam-Webster's Biographical Dictionary.*

A

absolute location a specific description of where a place is located; absolute location is often expressed using latitude and longitude (p. 12)
ubicación absoluta descripción específica del lugar donde se ubica un punto; con frecuencia se define en términos de latitud y longitud (pág. 12)

B

birthrate the annual number of births per 1,000 people (p. 88)
índice de natalidad número de nacimientos por cada 1,000 personas en un año (pág. 88)

C

cartography the science of making maps (p. 19)
cartografía ciencia de crear mapas (pág. 19)
climate a region's average weather conditions over a long period of time (p. 50)
clima condiciones del tiempo promedio de una región durante un período largo de tiempo (pág. 50)
command economy an economic system in which the central government makes all economic decisions (p. 94)
economía autoritaria sistema económico en el que el gobierno central toma todas las decisiones económicas (pág. 94)

communism a political system in which the government owns all property and dominates all aspects of life in a country (p. 92)
   comunismo sistema político en el que el gobierno es dueño de toda la propiedad y controla todos los aspectos de la vida de un país (pág. 92)

continent a large landmass that is part of Earth's crust; geographers identify seven continents (p. 36)
   continente gran masa de tierra que forma parte de la corteza terrestre; los geógrafos identifican siete continentes (pág. 36)

cultural diffusion the spread of culture traits from one region to another (p. 85)
   difusión cultural difusión de rasgos culturales de una región a otra (pág. 85)

cultural diversity having a variety of cultures in the same area (p. 83)
   diversidad cultural existencia de una variedad de culturas en la misma zona (pág. 83)

culture the set of beliefs, values, and practices that a group of people have in common (p. 80)
   cultura conjunto de creencias, valores y costumbres compartidas por un grupo de personas (pág. 80)

culture region an area in which people have many shared culture traits (p. 82)
   región cultural región en la que las personas comparten muchos rasgos culturales (pág. 82)

culture trait an activity or behavior in which people often take part (p. 81)
   rasgo cultural actividad o conducta frecuente de las personas (pág. 81)

deforestation the clearing of trees (p. 69)
   deforestación tala de árboles (pág. 69)

democracy a form of government in which the people elect leaders and rule by majority (p. 91)
   democracia sistema de gobierno en el que el pueblo elige a sus líderes y gobierna por mayoría (pág. 91)

desertification the spread of desertlike conditions (p. 65)
   desertificación ampliación de condiciones desérticas (pág. 65)

developed countries countries with strong economies and a high quality of life (p. 95)
   países desarrollados países con economías sólidas y una alta calidad de vida (pág. 95)

developing countries countries with less productive economies and a lower quality of life (p. 95)
   países en vías de desarrollo países con economías menos productivas y una menor calidad de vida (pág. 95)

earthquake a sudden, violent movement of Earth's crust (p. 38)
   terremoto movimiento repentino y violento de la corteza terrestre (pág. 38)

ecosystem a group of plants and animals that depend on each other for survival, and the environment in which they live (p. 63)
   ecosistema grupo de plantas y animales que dependen unos de otros para sobrevivir, y el ambiente en el que estos viven (pág. 63)

environment the land, water, climate, plants, and animals of an area; surroundings (pp. 12, 62)
   ambiente la tierra, el agua, el clima, las plantas y los animales de una zona; los alrededores (págs. 12, 62)

erosion the movement of sediment from one location to another (p. 39)
   erosión movimiento de sedimentos de un lugar a otro (pág. 39)

ENGLISH AND SPANISH GLOSSARY

ethnic group a group of people who share a common culture and ancestry (p. 83)
grupo étnico grupo de personas que comparten una cultura y una ascendencia (pág. 83)

extinct no longer here; a species that has died out has become extinct (p. 64)
extinto que ya no existe; una especie que ha desaparecido está extinta (pág. 64)

F

fossil fuels nonrenewable resources that formed from the remains of ancient plants and animals; coal, petroleum, and natural gas are all fossil fuels (p. 69)
combustibles fósiles recursos no renovables formados a partir de restos de plantas y animales antiguos; el carbón, el petróleo y el gas natural son combustibles fósiles (pág. 69)

freshwater water that is not salty; it makes up only about 3 percent of our total water supply (p. 31)
agua dulce agua que no es salada; representa sólo alrededor del 3 por ciento de nuestro suministro total de agua (pág. 31)

front the place where two air masses of different temperatures or moisture content meet (p. 53)
frente lugar en el que se encuentran dos masas de aire con diferente temperatura o humedad (pág. 53)

G

geography the study of the world, its people, and the landscapes they create (p. 4)
geografía estudio del mundo, de sus habitantes y de los paisajes creados por el ser humano (pág. 4)

glacier a large area of slow-moving ice (p. 31)
glaciar gran bloque de hielo que avanza con lentitud (pág. 31)

globalization the process in which countries are increasingly linked to each other through culture and trade (p. 97)
globalización proceso por el cual los países se encuentran cada vez más interconectados a través de la cultura y el comercio (pág. 97)

globe a spherical, or ball-shaped, model of the entire planet (p. 8)
globo terráqueo modelo esférico, o en forma de bola, de todo el planeta (pág. 8)

groundwater water found below Earth's surface (p. 32)
agua subterránea agua que se encuentra debajo de la superficie de la Tierra (pág. 32)

gross domestic product (GDP) the value of all goods and services produced within a country in a single year (p. 95)
producto interior bruto (PIB) valor de todos los bienes y servicios producidos en un país durante un año (pág. 95)

H

habitat the place where a plant or animal lives (p. 64)
hábitat lugar en el que vive una planta o animal (pág. 64)

human geography the study of the world's people, communities, and landscapes (p. 18)
geografía humana estudio de los habitantes, las comunidades y los paisajes del mundo (pág. 18)

humanitarian aid assistance to people in distress (p. 100)
ayuda humanitaria ayuda a personas en peligro (pág. 100)

humus (HYOO-muhs) decayed plant or animal matter; it helps soil support abundant plant life (p. 65)
  humus materia animal o vegetal descompuesta; contribuye a que crezca una gran cantidad de plantas en el suelo (pág. 65)

hydroelectric power the production of electricity from waterpower, such as from running water (p. 70)
  energía hidroeléctrica producción de electricidad generada por la energía del agua, como la del agua corriente (pág. 70)

interdependence a relationship between countries in which they rely on one another for resources, goods, or services (p. 99)
  interdependencia una relación entre países en que dependen uno de otros para obtener recursos, bienes, o servicios (pág. 99)

L

landform a shape on the planet's surface, such as a mountain, valley, plain, island, or peninsula (p. 35)
  accidente geográfico forma de la superficie terrestre, como una montaña, un valle, una llanura, una isla o una península (pág. 35)

landscape all the human and physical features that make a place unique (p. 4)
  paisaje todas las características humanas y físicas que hacen que un lugar sea único (pág. 4)

latitude the distance north or south of Earth's equator (p. 27)
  latitud distancia hacia el norte o el sur desde el ecuador (pág. 27)

lava magma that reaches Earth's surface (p. 37)
  lava magma que llega a la superficie terrestre (pág. 37)

map a flat drawing that shows all or part of Earth's surface (p. 8)
  mapa representación plana que muestra total o parcialmente la superficie de la Tierra (pág. 8)

market economy an economic system based on free trade and competition (p. 94)
  economía de mercado sistema económico basado en el libre comercio y la competencia (pág. 94)

meteorology the study of weather and what causes it (p. 20)
  meteorología estudio de las condiciones del tiempo y sus causas (pág. 20)

migration the movement of people from one place to live in another (p. 89)
  migración movimiento de personas de un lugar para ir a vivir a otro lugar (pág. 89)

monsoon a seasonal wind that brings either dry or moist air (p. 58)
  monzón viento estacional que trae aire seco o húmedo (pág. 58)

natural resource any material in nature that people use and value (p. 68)
  recurso natural todo material de la naturaleza que las personas utilizan y valoran (pág. 68)

nonrenewable resource a resource that cannot be replaced naturally; coal and petroleum are examples of nonrenewable resources (p. 69)
  recurso no renovable recurso que no puede reemplazarse naturalmente; el carbón y el petróleo son ejemplos de recursos no renovables (pág. 69)

ENGLISH AND SPANISH GLOSSARY

ENGLISH AND SPANISH GLOSSARY

ocean currents large streams of surface sea-water; they move heat around Earth (p. 52)
corrientes oceánicas grandes corrientes de agua de mar que fluyen en la superficie del océano; transportan calor por toda la Tierra (pág. 52)

permafrost permanently frozen layers of soil (p. 61)
permafrost capas de tierra congeladas permanentemente (pág. 61)

physical geography the study of the world's physical features—its landforms, bodies of water, climates, soils, and plants (p. 16)
geografía física estudio de las características físicas de la Tierra: sus accidentes geográficos, sus masas de agua, sus climas, sus suelos y sus plantas (pág. 16)

plate tectonics a theory suggesting that Earth's surface is divided into a dozen or so slow-moving plates, or pieces of Earth's crust (p. 36)
tectónica de placas teoría que sugiere que la superficie terrestre está dividida en unas doce placas, o fragmentos de corteza terrestre, que se mueven lentamente (pág. 36)

popular culture culture traits that are well known and widely accepted (p. 98)
cultura popular rasgos culturales conocidos y de gran aceptación (pág. 98)

population the total number of people in a given area (p. 86)
población número total de personas en una zona determinada (pág. 86)

population density a measure of the number of people living in an area (p. 86)
densidad de población medida del número de personas que viven en una zona (pág. 86)

precipitation water that falls to Earth's surface as rain, snow, sleet, or hail (p. 31)
precipitación agua que cae a la superficie de la Tierra en forma de lluvia, nieve, agua-nieve o granizo (pág. 31)

prevailing winds winds that blow in the same direction over large areas of Earth (p. 51)
vientos preponderantes vientos que soplan en la misma dirección sobre grandes zonas de la Tierra (pág. 51)

reforestation planting trees to replace lost forestland (p. 69)
reforestación siembra de árboles para reemplazar los bosques que han desaparecido (pág. 69)

region a part of the world that has one or more common features that distinguish it from surrounding areas (p. 6)
región parte del mundo que tiene una o más características comunes que la distinguen de las áreas que la rodean (pág. 6)

relative location a general description of where a place is located; a place's relative location is often expressed in relation to something else (p. 12)
ubicación relativa descripción general de la posición de un lugar; la ubicación relativa de un lugar suele expresarse en relación con otra cosa (pág. 12)

renewable resource a resource that Earth replaces naturally, such as water, soil, trees, plants, and animals (p. 69)
recurso renovable recurso que la Tierra reemplaza por procesos naturales, como el agua, el suelo, los árboles, las plantas y los animales (pág. 69)

revolution the 365¼-day trip Earth takes around the sun each year (p. 27)
revolución viaje de 365¼ días que la Tierra hace alrededor del Sol cada año (pág. 27)

rotation one complete spin of Earth on its axis; each rotation takes about 24 hours (p. 26)

rotación giro completo de la Tierra sobre su propio eje; cada rotación toma 24 horas (pág. 26)

savanna an area of tall grasses and scattered trees and shrubs (p. 58)

sabana zona de pastos altos con arbustos y árboles dispersos (pág. 58)

social science a field that focuses on people and the relationships among them (p. 5)

ciencias sociales campo de estudio que se enfoca en las personas y en las relaciones entre ellas (pág. 5)

solar energy energy from the sun (p. 26)

energía solar energía del Sol (pág. 26)

steppe a semidry grassland or prairie; steppes often border deserts (p. 59)

estepa pradera semiárida; las estepas suelen encontrarse en el límite de los desiertos (pág. 59)

surface water water that is found in Earth's streams, rivers, and lakes (p. 31)

agua superficial agua que se encuentra en los arroyos, ríos y lagos de la Tierra (pág. 31)

tropics regions close to the equator (p. 29)

trópicos regiones cercanas al ecuador (pág. 29)

United Nations an organization of countries that promotes peace and security around the world (p. 99)

Naciones Unidas organización de países que promueve la paz y la seguridad en todo el mundo (pág. 99)

![W]

water cycle the movement of water from Earth's surface to the atmosphere and back (p. 33)

ciclo del agua circulación del agua desde la superficie de la Tierra hacia la atmósfera y de regreso a la Tierra (pág. 33)

water vapor water occurring in the air as an invisible gas (p. 32)

vapor de agua agua que se encuentra en el aire en estado gaseoso e invisible (pág. 32)

weather the short-term changes in the air for a given place and time (p. 50)

tiempo cambios a corto plazo en la atmósfera en un momento y lugar determinados (pág. 50)

weathering the process by which rock is broken down into smaller pieces (p. 39)

meteorización proceso de desintegración de las rocas en pedazos pequeños (pág. 39)

ENGLISH AND SPANISH GLOSSARY

Economics Handbook

What Is Economics?

Economics may sound dull, but it touches almost every part of your life. Here are some examples of the kinds of economic choices you may have made yourself:

- Which pair of shoes to buy—the ones on sale or the ones you really like, which cost much more
- Whether to continue saving your money for the DVD player you want or use some of it now to go to a movie
- Whether to give some money to a fundraiser for a new park or to housing for the homeless

As these examples show, we can think of economics as a study of choices. These choices are the ones people make to satisfy their needs or their desires.

Glossary of Economic Terms

Here are some of the words we use to talk about economics:

ECONOMIC SYSTEMS

capitalism See market economy.

command economy an economic system in which the central government makes all economic decisions

communism a political system in which the government owns all property and runs a command economy

free enterprise a system in which businesses operate with little government involvement, as in a market economy

market economy an economic system based on free trade and competition; the government has little to say about what, how, or for whom goods are produced

mixed economy an economy that is a combination of command, market, and traditional economies

traditional economy an economy in which production is based on customs and tradition

THE ECONOMY AND MONEY

consumer one who buys goods or services for personal use

currency paper or coins that a country uses for its money supply

demand the amount of goods and services that consumers are willing and able to buy at a given time

economy the structure of economic life in a country

investment the purchase of something with the expectation that it will gain in value; usually property, stocks, etc.

productivity the amount of goods or services that a worker or workers can produce within a given amount of time

standard of living how well people are living; determined by the amount of goods and services they can afford

INTERNATIONAL TRADE

comparative advantage the ability of a company or country to produce something at a lower cost than other companies or countries

competition rivalry between businesses selling similar goods or services; a condition that often leads to lower prices or improved products

exports goods or services that a country sells and sends to other countries

free trade trade among nations that is not affected by financial or legal barriers; trade without barriers

gross domestic product (GDP) total market value of all goods and services produced in a country in a given year; per capita GDP is the average value of goods and services produced per person in a country in a given year

imports goods or services that a country brings in or purchases from another country

interdependence a relationship between countries in which they rely on one another for resources, goods, or services

market clearing price the price of a good or service at which supply equals demand

opportunity cost a trade-off; the value lost when producing or consuming one thing prevents producing or consuming another

profit the gain or excess made by selling goods or services over their costs

scarcity a condition of limited resources and unlimited wants by people

specialization a focus on only one or two aspects of production in order to produce a product more quickly and cheaply; for example, one worker washes the wheels of the car, another cleans the interior, and another washes the body

supply the amount of goods and services that are available at a given time

trade barriers financial or legal limitations to trade; prevention of free trade

PERSONAL ECONOMICS

barter the exchange of one good or service for another

credit a system that allows consumers to pay for goods and services over time

income a gain of money that comes typically from labor or capital

interest the money that a borrower pays to a lender in return for a loan

money any item, usually coins or paper currency, that is used in payment for goods or services

savings money or income that is not used to purchase goods or services

tax a required payment to a local, state, or national government; different kinds of taxes include sales taxes, income taxes, and property taxes

wage the payment a worker receives for his or her labor

RESOURCES

capital generally refers to wealth, in particular wealth that can be used to finance the production of goods or services

goods objects or materials that humans can purchase to satisfy their wants and needs

human capital sometimes used to refer to human skills and education that affect the production of goods and services in a company or country

natural resource any material in nature that people use and value

producer a person or group that makes goods or provides services to satisfy consumers' wants and needs

services any activities that are performed for a fee

Activities

1. On a separate sheet of paper, fill in the blanks in the following sentences:

    A. A _____ is an economic trade-off. If you produce or buy one thing, you can't produce or buy another.

    B. A person who buys goods or services is a _____.

    C. If we have an unlimited demand for a _____, such as oil, and there is only so much oil in the ground, we have a condition called _____.

    D. In order to raise money for roads, schools, and other public works, a government may require people to pay a _____.

    E. When a company increases its _____, its workers are producing more goods and services with less cost.

2. With a partner, compare prices in two grocery stores. Create a chart showing the price of five items in the two stores. Also, figure the average price of the items in each store. How do you think the fact that the stores are near each other affects prices? How might prices be different if one store went out of business? How might the prices be different or similar if the United States had a command economy? Present what you have learned about prices and competition to your class.

Index

KEY TO INDEX

c = **chart**	*m* = **map**
f = **feature**	*p* = **photo**

A

absolute location, 12
Academic Words, H2, H4
Active Reading, H3
Adelaide, 88c
Afghanistan: as developing country, 94c, 95
Africa: Africa: Physical, 126m; Africa: Political, 126m; climate zones of, 57m; energy production, 70c; Malawi, 17p; Victoria Falls, 17p
African plate, 36p
air mass, 53
Algeria: features and landscape of, 5p
Amazon River: rain forest, 58
Americas, 7
Andes, 37
animals: in ecosystem, 63, 63f; as renewable resource, 69
Antarctica: glaciers, 39; latitude of, 27
Antarctic Circle: midnight sun, 29, 29p
Antarctic plate, 36p
Arctic Circle: midnight sun, 29, 29p
Arizona: Grand Canyon, 40; Horseshoe Bend, 40, 40p; rock formation, 3p
Asia: Asia: Physical, 124m; Asia: Political, 125m; climate zones of, 57m; energy production, 70c; monsoons, 58
Atlantic Ocean: hurricanes, 53; Panama Canal, 40
Atlas: Africa: Physical, 126m; Africa: Political, 126m; Asia: Physical, 124m; Asia: Political, 125m; Europe: Physical, 122m; Europe: Political, 123m; North America: Physical, 118m; North America: Political, 119m; The North Pole, 129m; The Pacific: Political, 128m; South America: Physical, 120m; South America: Political, 121m; The South Pole, 129m; United States: Physical, 110–11m; United States: Political, 112–13m; World: Physical, 114–15m; World: Political, 116–17m
Australia: Adelaide, 88c; climate zones of, 57m; as developed country, 94c; population density of, 87
Australian plate, 36p
axis of Earth, 26–27

B

Bangladesh: floods, 33; rate of natural increase, 90
bar graph: analyzing, 74
baseball: cultural diffusion of, 84m, 85
bay: defined, H14
Biography: Eratosthenes, 18; Maathai, Wangari, 69; Wegener, Alfred, 37
birthrate: population changes and, 88–89; in Russia, 88
blizzards, 53
Brazil: Rio de Janeiro, 41p

C

California: Mediterranean climate, 59
Cameroon, 103c
Canada: ethnic group conflict, 83; government in, 91; rate of natural increase, 89–90
canyon: defined, H15
Canyonlands National Park, 39p
capitalism, 94
Caribbean South America: political map of, H12
cartography, 19; computer mapping, 19, 19m
Case Study: Ring of Fire, 42–43f
Chart and Graph Skills: Analyzing a Bar Graph, 74

Charter of the United Nations, The 100
charts and graphs: Average Annual Precipitation by Climate Region, 74c; Climate Graph for Nice, France, 59c; A Developed and a Developing Country, 102c; Developed and Less Developed Countries, 132c; Earth Facts, 130c; Economic Activity, 93c; Eruptions in the Ring of Fire, 42c; Essential Elements and Geography Standards, 13c; Irish Migration to the United States, 1845–1855, 89c; Mediterranean Climate, 59c; Percentage of Students on High School Soccer Teams by Region, 9c; Top Five Aluminum Producers, 2000, 77c; World Climate Regions, 56c; World Energy Production Today, 70c; World Languages, 133c; World Population Growth, 90c, 132c; World Religions, 133c
China: food/eating habits, 81; government of, 92
Chinatown, San Francisco, 7
climate, 48p, 75p; Average Annual Precipitation by Climate Region, 74c; defined, 50; dry, 59; highland, 60p, 61; large bodies of water, 53; limits on life and, 62; Mediterranean Climate, 59c; mountains and, 54, 54p; ocean currents, 52, 52m; overview of major climate zones, 55, 56–57c, 56–57m; polar, 61; rain shadow, 54, 54p; sun and location, 51; temperate, 59–60; tropical, 58; vs. weather, 50; wind and, 51–52
climate maps: West Africa, H13; World Climate Regions, 56–57m
coal: as nonrenewable energy resource, 69–70
coast: defined, H15
Colombia: Nevado del Ruiz, 42c
Colorado: mining industry, 40
Colorado River: erosion, 40, 40p
command economy, 94

INDEX

communism, 92
compass rose, H11
computers: computer mapping, 19, 19m
condensation, 32–33f, 33
conic projection, H9
coniferous trees, 60
continental drift, 36–37
continental plates, 36, 36p
continents, H7, 36; facts about, 131; movement of, 36–37
coral reef: defined, H14
core of Earth, 36
crust of Earth, 36
Cuba: economy of, 94; government of, 92
cultural diffusion, 84m, 85
cultural diversity, 83
cultural traits, 81
culture, 78p, 80–85; cultural diffusion, 84m, 85; cultural diversity, 83; cultural traits, 81; culture groups, 82–83; culture regions, 82; defined, 80; development of, 81–82; globalization and popular culture, 97–98; how culture changes, 84; innovation and, 84; landforms and, 40; midnight sun, 29, 29p; popular culture, 98; Tuareg of Sahara, 58; as a way of life, 80–81
culture groups, 82–83
culture region, 82; Arab Culture Region, 82, 82m, 82p; Japan, 82; Mexico, 82
cylindrical projection, H8

Dallas: airport, 11p
dams, 70
daylight: Earth's rotation and, 26–27
death rate: population changes and, 88–89; Russia, 88
deciduous trees, 60
deforestation, 69
degrees: defined, H6
delta, 40; defined, H14
democracy, 91–92
demography. *See* population

desert climate, 59; average annual precipitation, 74c; overview of, 56c
desertification, 65
deserts: defined, H15
developed countries, 94c, 95, 103c, 132c
developing countries, 94c, 95, 103c, 132c
dictatorship, 92
Dinosaur Age, North America in, 66–67f
Doctors Without Borders, 100
dodos, 64
Doldrums, 51m
dry climate, 57p, 59; overview of, 56–57m, 56c
dune: defined, H15

Earth: axis, 26–27; facts about, 130c; mapping, H6–H7; movement of, 26–27; revolution of, 27; rotation of, 26–27; solar energy and movement of, 26–27; structure, 130, 130p; tilt and latitude, 27; water supply of, 30–32
earthquakes: defined, 38; plate tectonics and, 38, 38p; Ring of Fire, 38, 42–43f
Earth's plates, 36–38, 36m, 36p
Earth's surface: erosion, 39–40; forces below, 36–38; forces on, 39–40; landforms, 35; plate tectonics, 36–38, 36p; weathering, 39
East Asia: North China Plain, 88
Eastern Hemisphere: defined, H7; map of, H7
economic activity, 93–94, 93p
economic geography, 19
economic indicators, 95
economics, 93–95; command economy, 94; developed and developing countries, 94c, 95; economic activity, 93–94, 93p; economic indicators, 95; Economics Handbook, 142–43; global economy, 98–99f; landforms and, 40; market economy, 94; natural resourc-

es and wealth, 72; systems of, 94; traditional economy, 94
economic systems, 94
ecosystems, 63, 63f
electricity: from coal, 70; hydroelectric energy, 70
employment: landforms and, 40
energy: from fossil fuels, 69–70; nonrenewable energy resources, 69–70; nuclear, 71; renewable energy resources, 70–71; solar, 27, 71; from water, 34, 70; from wind, 70; World Energy Production Today, 70c
Environment and Society, 13–14, 13c
environments, 12, 49p, 75p; changes to, 64; defined, 62; development of culture and, 81–82; Earth's changing environments, 66–67f; ecosystems, 63, 63f; limits on life, 62; soil and, 64–65
equator: climate at, 58; defined, H6; prevailing wind, 51m, 52, 52m
Eratosthenes, 18
erosion, 39–40, 39p, 40p; glaciers, 39; water, 40, 40p; wind, 39, 39p
essential elements of geography, H16–H17, 13–14
Ethiopia: food/eating habits, 81
ethnic group, 83
Euphrates: urban civilization and, 40
Eurasian plate, 36p, 37
Europe: climate zones of, 57m; energy production, 70c; Europe: Physical, 122m; Europe: Political, 123m; food/eating habits, 81
extinct, 64

fall, 29
five themes of geography, H16–H17, 10–12, 11f
flat-plane projection, H9
floodplain, 40
floods, 33, 53p

food/eating habits: China, 81; as cultural trait, 81; Ethiopia, 81; Europe, 81; Japan and Kenya, 81p; water and, 33–34
forest ecosystem, 63, 63f
forests: defined, H14; deforestation, 69; reforestation, 69
fossil fuels, 69–70
France: Mediterranean climate, 59; rate of natural increase, 90
freshwater, 31–32, 31p
front, 53

G

Geographic Dictionary, H14–H15
geography: defined, 4; economic, 19; essential elements, 13–14, H16–H17; five themes of, H16–H17, 10–12, 11f; global level, 7, 7p; human, 3, 16p, 18, 21; human-environment interaction theme, 10–12, 11f; landscape, 4; local level, 6, 6p; location theme, 10–12, 11f; movement theme, 10–12, 11f; physical, 3, 16–17, 16p, 21; place theme, 10–12, 11f; regional level, 6–7, 7p; region theme, 10–12, 11f; as science, 5; as social science, 5; studying, 4–9; urban, 19
Geography and History: Earth's Changing Environments, 66–67f
Geography and Map Skills Handbook: Geographic Dictionary, H14–H15; Geography Themes and Elements, H16–H17; Map Essentials, H10–H11; Map-making, H8–H9; Mapping the Earth, H6–H7; Working with Maps, H12–H13
Geography for Life, 14
Geography Skills: Analyzing Satellite Images, 15; Using a Physical Map, 44
geothermal energy, 70
Germany: as developed country, 95
glaciers, 31; defined, H15; erosion by, 39

globalization, 79p, 97–100; defined, 97; global economy, 98–99f; global trade, 99; interdependence and, 99; popular culture and, 97–98; world community, 99–100
global wind system, 51–52, 51m
globe, H6; as geographer's tool, 8–9
government: communism, 92; democracy, 91–92; dictatorship, 92; monarchy, 92; types of, 91–92; world governments, 92m
Grand Canyon, erosion and, 40
graphs. *See* charts and graphs
grasslands, 60
Great Lakes, 31; temperature in Michigan, 53
Great Plains, 60
Great Salt Lake, 31
Green Belt Movement, 69
Greenland: glaciers, 39
grid, H6
gross domestic product (GDP), 95
groundwater, 32; water cycle, 33
gulf: defined, H14
Gulf Stream, 52, 52m

H

habitat, 64
Hawaii: latitude of, 27
heat: weathering, 39
hemispheres: defined, H7
higher latitudes, 51
highland climate, 60p, 61; overview of, 56–57m, 57c
hill: defined, H15
Himalayas: plate tectonics and, 37, 38p
history: development of culture and, 81; Earth's Changing Environments, 66–67f
Horseshoe Bend, Arizona, 40, 40p
human-environment interaction theme, 10–12, 11f
human geography, 16p, 18, 21; defined, 3, 18; facts about, 132–33; uses of, 18
humanitarian aid, 100

Human Systems, 13–14, 13c
humid continental climate, 56c, 60
humid subtropical climate, 56c, 60
humid tropical climate, 56c, 58
humus, 64, 65
hurricanes, 53–54
Hussein, Saddam, 92
hydroelectric energy, 70
hydrology, 20

I

ice: erosion by glaciers, 39; as weathering, 39
ice cap climate, 61; overview of, 57c
Iceland: plate tectonics and, 37, 38p
immigrants: new cultural traits, 81
India: Mawsynram, 58; monsoons, 29; physical map of, H13, 44
Indian plate, 36p, 37
Indonesia: Krakatau, 42c; Tambora, 42c
infrared images, 15, 15p
innovation: cultural changes, 84
interdependence, 99
Internet Activities: Experiencing Extremes, 76; Researching Earth's Seasons, 46; Using Maps, 22; Writing a Report, 102
Iraq: government of, 92
Irish migration, 89, 89c, 89p
islands: defined, H7, H14, 35
isthmus: defined, H14
Italy: satellite images of, 15p

J

Japan: culture region, 82; eating habits, 81p; population density of, 87; rate of natural increase, 88

Kenya: eating habits, 81p; Nairobi, 88c; reforestation, 69p
Krakatau, 42c
Kurds, 82

L

lakes, 31; defined, H14
landforms, 45p; defined, 35; erosion of, 39–40; examples of, H14–H15; influence on life, 40–41; types of, 35; weathering of, 39
landscape: of Algeria, 5p; defined, 4
language: as cultural trait, 81; English as global language, 98; landforms and, 40; of the world, 133c
latitude: defined, H6, 27; solar energy and, 27
lava, 37
laws, 81
legend, H11
Lima, 88c
Literature: *River, The* (Gary Paulsen), 73
location: absolute, 12; relative, 12
location theme, 10–12, 11f; defined, 12
locator map, H11
London, 6p, 7p
longitude: defined, H6
lower latitudes, 51

M

Maathai, Wangari, 69
magma, 37
Malawi, 17p
Mali: rate of natural increase, 88–89
mantle of Earth, 36
manufacturing, 93, 93p
map projections, H8–H9; conic projection, H9; cylindrical projection, H8; flat-plane projection, H9; Mercator projection, H8

maps. *See also* Map Skills
advantages/disadvantages of, 8–9
Arab Culture Region, 82m
climate maps: West Africa, H13; World Climate Regions, 56–57m
compass rose, H11
Computer Mapping, 19, 19m
Conic Projections, H9
Cultural Diffusion of Baseball, 84m, 85
defined, H8, 8
Earth's plates, 36m
Eastern Hemisphere, H7
The First Crusade, 1096, H9
Flat-plane Projections, H9
as geographer's tool, 8–9
Global Wind System, 51m
High School Soccer Participation, 8m
Map Activity, 22, 46, 76, 87, 102
Mercator Projection, H8
Nice, France, 59m
Northern Hemisphere, H7
Pangaea, 66m
physical maps, 46m; Africa: Physical, 126m; Asia: Physical, 124m; Europe: Physical, 122m; India, 44; The Indian Subcontinent: Physical, H13; North America: Physical, 118m; The North Pole, 129m; South America: Physical, 120m; The South Pole, 129m; United States: Physical, 110–11m; World: Physical, 114–15m
political maps: Africa: Political, 126m; Asia: Political, 125m; Caribbean South America: Political, H12; Europe: Political, 123m; North America: Political, 119m; The Pacific: Political, 128m; South America: Political, 121m; United States: Political, 112–13m; World: Political, 116–17m
population maps: World Population Density, 87m
Ring of Fire, 42m
Southern Hemisphere, H7
Western Hemisphere, H7

World Governments, 92m
The World's Major Ocean Currents, 52m
Map Skills: Analyzing Satellite Images, 15; Interpreting Maps, 52m, 57m, 87; Understanding Map Projections, H8–H9; Using Different Kinds of Maps, H12–H13; Using Latitude and Longitude, H6–H7; Using a Physical Map, 44
Mariana Trench, 37
marine west coast climate, 56c, 60
market economy, 94
math connection: Calculating Population Density, 88
Mauritius, 64
Mawsynram, India, 58
Mediterranean climate, 59, 59p; Average Annual Precipitation, 74c; overview of, 56c
Mercator projection, H8
meridians: defined, H6
meteorology, 20
Mexico: culture region, 82; government in, 91
Michigan: Great Lakes and temperature, 53
Mid-Atlantic Ridge, 37
Middle East: energy production, 70c
midnight sun, 29, 29p
migration, 89; Irish, 89, 89c, 89p
mineral resources, 71
minutes: defined, H6
Mississippi River: delta, 40; tributaries of, 31
Missouri River: as tributary of Mississippi River, 31
Mojave Desert, 11p
monarchy, 92
monsoons, 58; India, 29
mountains: climate, 54, 54p; defined, H15, 35; glaciers and, 39; plate tectonics and, 37, 38p; rain shadow, 54, 54p
Mount Pinatubo, 42c, 43
Mount Rainier, 11p
Mount Saint Helens: Ring of Fire, 42–43f
movement: of Earth and solar energy, 26–27; migration, 89; plate tectonics, 36–38
movement theme of geography, 10–12, 11f
mudflow, 43

N

Nairobi, 88c
natural environments. *See* environments
natural gas: as nonrenewable energy resource, 69–70
natural resources, 49p, 68–72, 75p; defined, 68; economic activity and, 93, 93p; energy resources, 69–71; managing, 69; mineral resources, 71; nonrenewable, 69; nonrenewable energy resources, 69–70; nuclear energy, 71; people using in daily life, 71; renewable, 69; renewable energy resources, 70–71; Top Five Aluminum Producers, 2000, 77c; types of, 69; wealth and, 72
Nazca plate, 36, 36p, 37
Nepal, 3p; as developing country, 95
Nevado del Ruiz, Colombia, 42c
New Guinea: language and landforms, 40
New York City: cultural diffusion of baseball, 84m, 85
Nice, France: Mediterranean climate, 59
Nigeria: as developing country, 95; rate of natural increase, 90
night: Earth's rotation and, 26–27
Nile River: delta, 40
nonrenewable energy resources, 69–70
nonrenewable resources, 69
North Africa: Arab culture region, 82, 82m, 82p
North America: climate zones of, 56m; in Dinosaur Age, 66–67f; energy production, 70c; North America: Physical, 118m; North America: Political, 119m
North American plate, 36p, 37
North Atlantic Drift, 52, 52m
North China Plain, 88
Northern Hemisphere: defined, H7; map of, H7; seasons in, 28–29

North Korea: economy of, 94
North Pole: 129m; climate, 61; prevailing winds, 52;
Norway: government of, 92
note-taking skills, 4, 10, 16, 26, 30, 35, 50, 55, 62, 68, 86, 91, 97
nuclear energy, 71

O

oasis: defined, H15
ocean currents, 52, 52m; world's major, 52m
ocean plates, 36, 36p
oceans: defined, H14; erosion, 40; salt water, 31
ocean trenches: plate tectonics and, 37
oil. *See* petroleum
Olympics, 79p

P

Pacific Ocean: Panama Canal, 40; Ring of Fire, 42–43f; typhoons, 53
Pacific plate, 36p, 37
Panama Canal, 40
Pangaea, 66f, 66m
parallels: defined, H6
peninsula, 35; defined, H14
per capita GDP, 95
permafrost, 61
Peru: Lima, 88c
petroleum: as nonrenewable energy resources, 69–70; products from, 72p
Philippines: Mount Pinatubo, 42c, 43
physical features: and climate, 54; in geography, 16–17; on the land, H14–15, 35–41
physical geography, 16–17, 16p, 21; defined, 3, 16; uses of, 17
physical maps, 46m; Africa: Physical, 126m; Asia: Physical, 124m; defined, H13; Europe: Physical, 122m; India, 44; The Indian Subcontinent:

Physical, H13; North America: Physical, 118m; The North Pole, 129m; South America: Physical, 120m; The South Pole, 129m; United States: Physical, 110–11m; using, 44; World: Physical, 114–15m
Physical Systems, 13–14, 13c
Places and Regions, 13–14, 13c
place theme, 10–12, 11f
plains, 35; defined, H15
plants: in ecosystem, 63, 63f; as renewable resource, 69; soil and, 64–65
plateau: defined, H15
plate tectonics, 36–38, 36p, 130; continental drift, 36–37; erosion, 39–40, 39p, 40p; mountains, 37; ocean trenches, 37; plates collide, 37; plates separate, 37, 38p; plates slide, 38, 38p
polar climate, 57p, 61; overview of, 56–57m, 57c
Polar easterlies, 51m, 53
political maps: Africa: Political, 126m; Asia: Political, 125m; Caribbean South America: Political, H12; defined, H12; Europe: Political, 123m; North America: Political, 119m; The Pacific: Political, 128m; South America: Political, 121m; United States: Political, 112–13m; World: Political, 116–17m
pollution: water, 33
popular culture: defined, 98; globalization and, 98
population, 79p, 86–90; birthrate, 88; death rate, 88–89; defined, 86; migration, 89; population density, 86–87; rate of natural increase, 88; tracking population changes, 88–89; world population growth, 89–90, 90c; world trends in, 89–90, 132, 132c
population density, 86–87; of Australia, 87; calculating, 88; defined, 87; of Japan, 87; where people live and, 87–88; world population density, 87m

population maps: World Population Density, 87m
precipitation: Average Annual Precipitation by Climate Region, 74c; defined, 31; mountains and, 54; water cycle and, 32–33f, 33
predators: in ecosystem, 63f
prevailing winds, 51–52, 51m
primary industry, 93, 93p

Primary Sources: *Charter of the United Nations, The,* 100; *Geography for Life,* 14; *Time Enough for Love,* 50
prime meridian, H6
projections. *See* map projections

quaternary industry, 93p, 94
Quick Facts: Climate, Environment, and Resources visual summary, 75; A Geographer's World visual summary, 21; Planet Earth visual summary, 45; The World's People visual summary, 101

rainfall: limits on life and, 62; seasons and, 29; tropics, 29
rain forests, 58; clearing, 64
rain shadow, 54, 54p
rate of natural increase, 88; Bangladesh, 90; Canada, 89–90; France, 90; Japan, 88; Mali, 88–89; Nigeria, 90
Reading Skills: Reading Social Studies, H1–H3; Understanding Cause and Effect, 48, 108; Understanding Main Ideas, 78, 109; Using Prior Knowledge, 2, 106; Using Word Parts, 24, 107
recreation: water and, 34
Red Cross, 100
reforestation, 69

region, 6–7
region theme, 10–12, 11f
relative location, 12
religion: development of culture and, 81; of the world, 133c
renewable energy resources, 70–71
renewable resources, 69
revolution of Earth, 27
Ring of Fire, 38, 42–43f
Rio de Janeiro, 41p
River, The **(Gary Paulsen),** 73
rivers, 31; defined, H14; erosion, 40, 40p
rock: in soil, 65p
rock formation: Arizona, 3p
Rocky Mountains, 6
Roman Catholic religion, 81
rotation of Earth, 26–27; prevailing winds, 52
runoff, 32–33f, 33
Russia: birthrate, 88; death rate, 88
Rwanda: ethnic group conflict, 83

Sahara, 6; culture and, 82; Tuareg, 58
salt water, 31, 31p
San Andreas Fault, 38
sand: wind erosion, 39, 39p
satellite images: analyzing, 15; as geographer's tool, 9; infrared images, 15, 15p; true color, 15, 15p
Saudi Arabia: government of, 92
savannas, 58
scale, map, H11
Scavenger Hunt, H20
science and technology: computer mapping, 19, 19m; soil factory, 64
seasons, 28–29; based on temperature and length of day, 28; midnight sun, 29, 29p; in Northern Hemisphere, 28; rainfall and, 29; solar energy and, 28–29; Southern Hemisphere, 28–29; spring and fall,

29; tilt of Earth and, 28–29; winter and summer, 28–29
secondary industry, 93, 93p
sediment, 40
Singapore, 103c
sinkhole: defined, H14
soccer: as cultural trait, 81
social science, 5
Social Studies Skills
    Chart and Graph Skills: Analyzing a Bar Graph, 74
    Geography Skills: Analyzing Satellite Images, 15; Using a Physical Map, 44
    Study Skills: Organizing Information, 96
Social Studies Words, H4
soil: environments, 64–65; humus, 64, 65; layers of, 65p; limits on life and, 62; losing fertility, 65; as renewable resource, 69; soil factory, 64
solar energy, 45p; defined, 27; in ecosystem, 63, 63f; movement of Earth and, 27; as renewable energy, 71; seasons and, 28–29; water cycle and, 32–33f, 33; weather and climate, 51
South America: Andes, 37; climate zones of, 56m; energy production, 70c; South America: Physical, 120m; South America: Political, 121m
South American plate, 36p, 37
Southeast Asia: language and landforms, 40; terrace farming, 40–41
Southern Hemisphere: defined, H7; map of, H7; seasons in, 28–29
South Pole: climate, 61; prevailing winds, 52; The Sorth Pole, 129m
Southwest Asia: Arab culture region, 82, 82m, 82p
Southwestern United States, 33
Spain: government of, 92
special-purpose maps, H13
spring, 29
springs: groundwater and, 32
Standardized Test Practice, 23, 47, 77, 103

steppe climate, 59; average annual precipitation, 74c; overview of, 56c
storm surges, 54
strait: defined, H14
stratovolcanoes, 43
streams, 31
Study Skills: Organizing Information, 96
subarctic climate, 57c, 61
subsoil, 65p
summer, 28–29
surface water, 31
Surtsey Island, 38p

taiga, 61
Tambora, 42c
Tanzania: ethnic groups of, 83
technology. *See* science and technology
temperate climate, 59–60; overview of, 56–57m, 56c
temperature: large bodies of water, 53; limits on life and, 62; mountains and, 54
tertiary industry, 93, 93p
thunderstorms, 48–49p, 53
Tigris River: urban civilization and, 40
tilt of Earth, 27; climate and weather, 51; seasons and, 28–29
topsoil, 65p
tornadoes, 53, 53p
trade: globalization and, 99
trade winds, 51m
traditional economy, 94
trees: Green Belt Movement, 69; as renewable resource, 69
tributary, 31
Tromso, Norway, 29
tropical climate, 56p, 58; overview of, 56–57m, 56c
tropical humid climate, 58; average annual precipitation, 74c
tropical savanna climate, 58; average annual precipitation, 74c; overview of, 56c
Tropic of Cancer, 58

Tropic of Capricorn, 58
tropics: defined, 29; rainfall and, 29
true color images, 15, 15p
tundra climate, 61; average annual precipitation, 74c; overview of, 57c
typhoons, 53

Ukraine, 103c
Understanding Cause and Effect, 48, 108
Understanding Main Ideas, 78, 109
United Nations Children's Fund (UNICEF), 100
United Nations (UN), 99; *Charter of the United Nations, The,* 100
United States, 23m; as developed country, 95; economic system of, 94; ethnic groups in, 83; global popular culture and, 98; government in, 91; High School Soccer Participation, 8m; United States: Physical, 110–11m; United States: Political, 112–13m; water shortages, 33
urban geography, 19
Uruguay, 103c
Uses of Geography, 13–14, 13c
U-shaped valleys, 39
Using Prior Knowledge, 2, 106
Using Word Parts, 24, 107
Utah: Great Salt Lake, 31

valley, 35; defined, H15
Victoria Falls, 17p
Viewing Skills: Creating and Viewing a Weather Report, 48
Vocabulary, H4
volcanoes: defined, H15; Ring of Fire, 38, 42–43f; stratovolcanoes, 43

Washington, D.C., 11p
water, 30–32, 45p; benefits of, 33–34; drought, 33; erosion, 40, 40p; flooding, 33; food and, 33–34; freshwater, 31–32, 31p; groundwater, 32; hydroelectric energy, 70; importance of, 30; limits on life and, 62; pollution, 33; as renewable resource, 69; salt, 31, 31p; surface, 31; in water cycle, 32–33, 32f; weathering, 39
water cycle, 32–33, 32–33f, 47p
water vapor, 32–33, 32–33f
wealth: natural resources and, 72
weather: blizzards, 53; vs. climate, 50; defined, 50; front, 53; hurricanes, 53–54; ocean currents, 52, 52m; storm surges, 54; sun and location, 51; thunderstorms, 53; tornadoes, 53, 53p; wind and, 51–52
weathering, 39
Wegener, Alfred, 37
West Africa: climate map of, H13
Westerlies, 51m, 52, 52m, 53
Western Hemisphere: defined, H7
wetland: defined, H14
wind: climate and, 51–52; erosion, 39, 39p; monsoons, 58; ocean currents, 52, 52m; prevailing, 51–52, 51m; as renewable energy, 70
winter, 28–29
world facts, 130–33
World in Spatial Terms, 13–14, 13c
Writing Skills: Creating a Poster, 78; Explaining a Process, 104; Writing a Haiku, 24; Writing a Job Description, 2

Yardley, Pennsylvania, 53p

INDEX

Credits and Acknowledgments

Acknowledgments

For permission to reproduce copyrighted material, grateful acknowledgment is made to the following sources:

National Geographic Society: From *Geography for Life: National Geography Standards 1994.* Copyright © 1994 by National Geographic Research & Exploration. All rights reserved.

G. P. Putnam's Sons, a division of Penguin Group (USA) Inc.: From *Time Enough for Love, the Lives of Lazarus Long* by Robert Heinlein. Copyright © 1973 by Robert Heinlein. All rights reserved.

Random House Children's Books, a division of Random House, Inc.: From *The River* by Gary Paulsen. Copyright © 1991 by Gary Paulsen.

United Nations: From the *Preamble to the Charter of the United Nations.* Copyright © 1945 by United Nations.

Sources used by The World Almanac® for charts and graphs:

Eruptions in the Ring of Fire: *The World Almanac and Book of Facts, 2005;* World Energy Production: Energy Information Administration of the U.S. Department of Energy; A Developed and a Developing Country: *The World Factbook, 2005;* U.S. Census Bureau; World Health Organization

Illustrations and Photo Credits

Frontmatter: ii, Victoria Davis/HRW; iv, Sharna Balfour/Gallo Images/Corbis; vi, Liu Liqun/Corbis; vii, Eric Meola/Getty Images; viii (br), Planetary Visions; xi, Tom Nebbia/Corbis; H16 (t), Earth Satellite Corporation/Science Photo Library; H16 (tc), Frans Lemmens/Getty Images; H16 (c), London Aerial Photo Library/Corbis; H16 (bc), Harvey Schwartz/Index Stock Imagery; H16 (b), Tom Nebbia/Corbis. Acetate Inserts: (t), Gavin Hellier/Robert Harding/Getty Images; (c), Rex Butcher/Jon Arnold Images; (b), Steve McCurry/Magnum Photos.

Introduction: A, Taxi/Getty Images; B (bl) Stephen Frink/Digital Vision/Getty Images; B (cr), Frans Lemmens/The Image Bank/Getty Images; 1 (bc), Robert Harding/Digital Vision/Getty Images; B–1 (background satellite photos), Planetary Visions.

Chapter 1: 2 (br), M. Colonel/Photo Researchers, Inc.; 2–3 (t), Age Fotostock/SuperStock; 3 (br), Anthony Cassidy/Getty Images; 3 (bl), Tom Bean/Corbis; 5 (b), Frans Lemmens/Getty Images; 6 (b), Kim Sayer/Corbis; 7 (bl), London Aerial Photo Library/Corbis; 7 (br), ESA/K.Horgan/Getty Images; 8 (tl), Michael Newman/PhotoEdit; 11 (tl), David R. Frazier/Photo Researchers, Inc.; 11 (tr), Richard Bryand/Arcaid Picture Library; 11 (bl), David Muench/Corbis; 11 (bc), AFP/Getty Images; 11 (br), Morton Beebe/Corbis; 12–13 (b), Tom Nebbia/Corbis; 15 (l), M-SAT Ltd./Science Photo Library; 15 (r), Earth Satellite Corporation/Science Photo Library; 17 (tl), Torleif Svensson/Corbis; 17 (tr), Penny Tweedie/Stone/Getty Images; 19 (b), Donna Cox and Robert Patterson/NCSA; 20 (tl), Joe Raedle/Getty Images; 21 (tl), Frans Lemmens/Getty Images; 21 (tc), Penny Tweedie/Stone/Getty Images; 21 (tr), Donna Cox and Robert Patterson/NCSA.

Chapter 2: 24 (br), Pete Saloutos/Corbis; 24–25 (t), Earth Satellite Corporation/Science Photo Library; 25 (br), Royalty Free/Corbis; 25 (bl), George H.H. Huey/Corbis; 29 (tr), Paul A. Souders/Corbis; 30–31 (b), Doug Wilson/Corbis; 31 (br), Terje Rakke/The Image Bank/Getty Images; 34 (tl), Rick Doyle/Corbis; 34 (tr), Alan Sirulnikoff/Photo Researchers, Inc.; 37 (bl), Bettmann/Corbis; 38 (cr), Yann Arthus-Bertrand/Corbis; 38 (br), Galen Rowell/Corbis; 39 (br), Age Fotostock/SuperStock; 40 (bl), Owaki-Kulla/Corbis; 41 (t), Age Fotostock/SuperStock; 43 (tl), David Weintraub/Photo Researchers, Inc.; 43 (tr), Gary Braasch/Corbis; 45 (tl), W.H. Mueller/Zefa Images/Corbis; 45 (tc), Rick Doyle/Corbis; 45 (tr), Galen Rowell/Corbis.

Chapter 3: 48 (br), Kate Thompson/National Geographic/Getty Images; 48–49 (t), Warren Faidley/WeatherStock; 49 (bl), L. Clarke/Corbis; 49 (br), Bill Ross/Corbis; 53 (tr), William Thomas Cain/Getty Images; 53 (br), Eric Meola/Getty Images; 56 (tl), Age FotoStock/SuperStock; 57 (tr), AlaskaStock; 57 (cr), Royalty-free/Corbis; 58 (cl), Martin Harvey/Corbis; 59 (br), Ingram/PictureQuest; 60 (t), Sharna Balfour/Gallo Images/Corbis; 64 (b), Carl and Ann Purcell/Corbis; 69 (tr), William Campbell/Peter Arnold, Inc.; 69 (c), Adrian Arbib/Corbis; 71 (tr), James L. Amos/Corbis; 71 (cr), Creatas/PictureQuest; 72 (tl), Sarah Leen/National Geographic Image Collection; 73 (tr), James Randklev/Getty Images; 75 (tl), Royalty-free/Corbis; 75 (tr), James L. Amos/Corbis.

Chapter 4: 101 (tl), Sebastian Bolesch/Peter Arnold, Inc.; 101 (tc), Rosenfeld Images Ltd./Photo Researchers, Inc.; 101 (tr), Reuters/Corbis; 78 (b), Liu Liqun/Corbis; 78–79 (t), Getty Images; 79 (br), Shahn Rowe/Stone/Getty Images; 79 (bl), Richard I'Anson/Lonely Planet Images; 81 (tl), Tom Wagner/Corbis; 81 (tr), Knut Mueller/Peter Arnold, Inc.; 83 (bl), Peter Armenia; 83 (br), Sebastian Bolesch/Peter Arnold, Inc.; 84 (cl), Reuters/Corbis; 84 (cr), Timothy A. Clary/AFP/Getty Images; 84–85 (t), Courtesy Library of Congress; 89 (t), The Granger Collection; 90 (cr), Marcel & Eva Malherbe/The Image Works; 90 (tr), Peter Beck/Corbis; 93 (t), Owaki-Kulla/Corbis; 93 (tc), Rosenfeld Images, Ltd./Photo Researchers, Inc.; 93 (bc), Kevin Fleming/Corbis; 93 (b), Michelle Garrett/Corbis; 94 (bl), Glen Allison/Stone/Getty Images; 94 (br), Carl & Ann Purcell/Corbis.

Backmatter: 131, Planetary Visions; 132–133 (t), Oriental Touch/Robert Harding; 133 (cl), Amanda Hall/Robert Harding; 142 (tr), Tom Stewart/Corbis.